# Economic Policy-Making in the Asia-Pacific Region

faciliter cette prise de conscience, c'est pourquoi il a entrepris, au cours des quelques dernières années, plusieurs initiatives bilatérales et multilatérales dans la région de l'Asie du Pacifique. La série d'ateliers de l'IRP sur "L'évolution des gouvernements de la région de l'Asie du Pacifique" est au centre même de ces initiatives. Cette série de rencontres a pour but de mettre en rapport les uns avec les autres des chercheurs de haut rang et des praticiens de politique générale des différents pays de la région, pour leur permettre d'échanger des idées sur la capacité de gouverner et sur la gestion publique des questions et des réformes et, grâce à la dissémination des comptes rendus d'ateliers, de toucher un vaste auditoire d'observateurs intéressés dans ces discussions. Le premier atelier s'est tenu à Victoria en juin 1987. Le second, pour lequel les communications réunies dans ce volume ont été préparées, était co-parrainé par le Thailand Development Research Institute (TDRI – Institut de développement de la recherche de Thaïlande); il a eu lieu à Bangkok en décembre 1988. Le troisième est prévu pour mai 1990 à Kuala Lumpur.

L'atelier de 1987, au cours duquel on a examiné l'évolution des statuts des institutions et des procédés gouvernementaux de la région, avait pour objectif de poser les fondements d'une discussion plus précise pour les rencontres suivantes. Étant donné le remarquable succès atteint par de nombreux pays de l'Asie du Pacifique en ce qui a trait à l'élaboration de stratégies économiques orientées vers l'extérieur, il n'est pas surprenant d'apprendre que le thème de l'atelier ait porté sur la manière dont les politiques économiques sont élaborées dans les pays de la région. Au cours de la rencontre de Kuala Lumpur, on examinera encore plus en détail ces questions et on se penchera tout particulièrement sur le rôle joué par les instituts politiques, encore appelés "cuves de réflexion", dans un monde de plus en plus globalisé et interdépendant.

L'Institut est heureux de publier ce second volume où sont réunis les communications présentées au cours de l'atelier. Bien qu'il n'offre aucun modèle standard pouvant produire des miracles économiques, la diversité des approches décrites permettra de réfléchir à tous ceux qui s'intéressent à la formulation des politiques économiques ou qui y prennent activement part. Nous espérons également qu'il contribuera à l'établissement de liens

# Avant-propos

Peu d'entre nous pourraient contester le fait que nous sommes entrés dans "l'âge du Pacifique". Les preuves sont irréfutables. Avec le Japon à leur tête, les pays de l'est et du sud-est asiatique ont, l'un après l'autre, gravi les barreaux de l'échelle du développement économique, et d'ici la fin du siècle, on estime que cette région sera le fournisseur de plus de 50 pour cent de la production globale et de 70 pour cent du commerce. Au même moment, nous parlons souvent, au Canada, de notre affiliation à cette communauté de l'Asie du Pacifique. Pourtant, bien qu'il soit vrai que les liens relatifs au commerce, aux investissements et à l'immigration lancés au travers du Pacifique soient devenus importants, il est également vrai que notre connaissance des forces sous-jacentes aux développements remarquables de cette région est loin d'être adéquate. Pour que le Canada puisse véritablement se réclamer du rôle vital qu'il joue dans cette communauté, et pour qu'il puisse bénéficier des avantages que celle-ci peut lui offrir, il doit s'efforcer davantage de comprendre les cultures, les traditions et les aspirations des habitants de cette région.

L'Institut a compris l'importance qu'il y avait d'établir des rapports personnels continus et des réseaux institutionnels pour is the IRPP workshop series on "The Changing Shape of Government in the Asia-Pacific Region". This series of meetings is intended to link senior public policy researchers and practitioners from around the region, providing a vehicle for the exchange of ideas on governance and public management issues and reforms, and, through the dissemination of the workshop proceedings, to involve a larger cadre of interested observers in these discussions. The first workshop was held in Victoria in June, 1987; the second, for which the papers in this volume were prepared, was co-hosted with the Thailand Development Research Institute (TDRI) in Bangkok in December, 1988; and a third is scheduled for May, 1990 in Kuala Lumpur.

The 1987 workshop, with its sweeping look at the status of government institutions and processes in the region, was designed to lay the foundation for a somewhat more focused discussion in subsequent meetings. Given the remarkable success that many of the Asia-Pacific countries have had in developing effective outward-oriented economic strategies, it is perhaps not surprising that the topic for the Bangkok workshop was how economic policy is made in the region. The Kuala Lumpur meeting will continue to sharpen the focus of the series and will examine the role that policy institutes or "think tanks" play in an increasingly globalized and interdependent world.

The Institute is pleased to release this second volume of workshop papers. Although it offers no blueprint for creating economic miracles, the diversity of approaches described offers much food for thought to all who are involved in or concerned with making economic policy. We also hope that it will contribute to the building of long-term institutional and personal linkages across the Pacific waters.

The Institute is also pleased to acknowledge the generous financial assistance given to this workshop series by the Management for Change Program of the Canadian International Development Agency, and the cooperation of the Thailand Development Research Institute in co-hosting the 1989 meeting.

Rod Dobell
President
January 1990

# Foreword

Few among us would argue the notion that we have entered the "Age of the Pacific". The evidence is overwhelming. Led by Japan, country after country in East and Southeast Asia has vaulted up the ladder of economic development, and by the end of this century it is estimated that the region will be home to over 50 percent of global production and 70 percent of trade. At the same time, we in Canada often speak of our membership in this Asia-Pacific community. But, while it is true that our trade, investment and immigration links across the Pacific have become very significant, it is also true that our knowledge of the forces driving the remarkable developments in the region is far from adequate. If Canada is to give force to its claim to be a vital part of this community, and to benefit from the opportunities therein, it must greatly intensify its efforts to understand the cultures, traditions and aspirations of the people of the region.

Recognizing the importance of establishing sustained personal relationships and institutional networks as a means of facilitating this increased awareness, the Institute has, over the last several years, entered into a number of bilateral and multilateral initiatives in the Asia-Pacific region. Central to this effort

## Indonesia
The Indonesian Deregulation Process:
Problems, Constraints and Prospects .................... 321
*Sjahrir*

## Related Publications ............................. 341

## New Zealand
The New Zealand Experience .......................... 79
*Gary Hawke*

## Philippines
A Political Economy of Philippine Policy-Making ........ 109
*Emmanuel S. de Dios*

## Japan
The Administrative Reform, Restructuring
and Economic Planning in Japan ...................... 149
*Yoichi Okita*

## Thailand
Economic Policy-Making in a Liberal
Technocratic Polity .................................... 181
*Chai-Anan Samudavanija*

## South Korea
Political Transition and Economic
Policy-Making in South Korea ......................... 203
*Chung-Si Ahn*

## China
The Process and the Problems of the
Decentralization of China's Economic
Decision-Making Power: 1978-1988 .................... 231
*Cao Yuan-zheng*

## Hong Kong
Economic Policy under Stress ......................... 251
*H.C. Kuan*

## Malaysia
The Shaping of Economic Policy in a
Multi-Ethnic Environment: The
Malaysian Experience ................................ 273
*Mavis Puthucheary*

## Singapore
Changing the Economic Policy-Making Process
in Singapore: Promise and Problems ................... 299
*Linda Low*

# Contents

**Foreword** .................................................... ix

**Avant-propos** ................................................ xi

**Introduction**
Economic Policy-Making in the Asia-Pacific
Nations: Processes and Problems ...................... 1
*John W. Langford and K. Lorne Brownsey*

**Australia**
Australian Economic Policy:
Problems and Processes ................................ 11
*Michael Keating and Geoff Dixon*

**Canada**
Economic Policy-Making in Canada:
The Case of the Canada-U.S. Free Trade
Agreement ............................................. 57
*A.R. Dobell*

Copyright © The Institute for Research on Public Policy 1990
All rights reserved

Printed in Canada

Legal Deposit First Quarter
Bibliothèque nationale du Québec

## Canadian Cataloguing in Publication Data

Main entry under title:
Economic policy-making in the Asia-Pacific region

Papers prepared for a workshop held in Bangkok,
Thailand, in December 1988.
Prefatory material in English and French.
ISBN 0-88645-104-3

1. Asia, Southeastern—Economic policy.
2. Pacific Area—Economic policy.
I. Langford, John W.
II. Brownsey, K. Lorne, 1952-
III. Institute for Research on Public Policy

HC412.E26 1990    330.9182'3    C90-097558-X

*Camera-ready copy and publication management by*
PDS Research Publishing Services Ltd.
P.O. Box 3296
Halifax, Nova Scotia B3J 3H7

*Published by*
The Institute for Research on Public Policy
L'Institut de recherches politiques
P.O. Box 3670 South
Halifax, Nova Scotia B3J 3K6

# Economic Policy-Making in the Asia-Pacific Region

Edited by

John W. Langford

and

K. Lorne Brownsey

The Institute for Research on Public Policy
L'Institut de recherches politiques

the lens of the recent decision to enter into a free trade agreement with the United States, however, Rod Dobell suggests that narrow, neo-conservative alliances between government and business around key economic decisions are becoming the rule rather than the exception. "The government's economic policy agenda over its first four years tended to parallel very closely that of the business community, primarily big business and multinational enterprises." This has resulted in a decision-making system that effectively shuts out large numbers of narrow interests without—as is the case in Australia—legitimizing any encompassing group through which their concerns can be effectively heard. Dobell and others see this half-a-loaf corporatism as representing a significant danger to the economic policy-making system in Canada. "The issue, at its briefest, is the need to find ways to involve more effectively the much broader range of economic and political players who will be necessary participants in formulating and implementing economic policy in the future. The relatively narrow band of players who could expect to carry the day in earlier times—senior federal ministers and officials, leaders of key Canadian corporations, and a small number of other opinion leaders—will no longer suffice."

Clearly Canada and Australia have moved some way from the customary pluralist conception of a government playing a stand-offish, brokerage or accommodation role among a large number of narrow pressure groups of relatively equal strength. The same cannot be said for New Zealand where government continues to act in the more traditional role as a broker of the interests of its various constituents. Although Gary Hawke argues that the real action in recent years has centered around the substance of policy rather than the way it is made, he records the government's efforts to make considerations of equity major factors in the reform of key economic policies. This effort has resulted in the widening of the scope of policy-making consultations, the introduction of the Cabinet Social Equity Committee designed to link economic and social policy-making, and the development of Working Groups to increase the breadth of expert opinion being presented to this Committee. This structural recognition of both the interrelationship of economic and social policy and the need to widen the participation in economic decision-making may have had only modest success, but it is at least an attempt to address the issues raised by Dobell in the Canadian context.

Signs of institutional ferment, however, are not limited to the western-type polities in the Asia-Pacific region. In a literal sense, quite revolutionary changes have taken place in the Philippines with the overthrow of Marcos and the accession to power of Corazon Aquino. One aspect of the turnover which deserves attention here is the degree to which the economic technocrats have been displaced from key positions in the Aquino administration by the appointment of prominent businessmen. However, as Emmanuel S. de Dios explains in his paper, it is hard to discern the significance of the replacement of Marcos's "crony capitalists" by the Aquino "economic elite" and a revived Congress dominated by traditional politicians.

Yoichi Okita uses three case studies to provide the reader with different perspectives on how the Japanese economic decision-making system can be characterized. His claim that the system can best be described as pluralist is strengthened by evidence of the erosion of bureaucratic (rational planning) power in Japan and the emergence of economic circumstances in which the government may have to give up the consensus building role (corporatism) and more often favour the interests of one group over another. If Okita is correct, "Japan Inc." and the relationship of mutual dependency and constant adjustment between politicians, bureaucrats, industrialists, and farmers, etc. may not be immutable features of the emerging economic policy-making process. However, the image of the pendulum which he introduces towards the end of his paper is intended to caution the reader against leaning too heavily on the pluralist model alone to try to capture the evolving character of the Japanese economic decision-making process.

Equally significant shifts in Thailand have started from the bureaucratic-authoritarian end of the spectrum of economic policy-making models. In recent years, the army and the bureaucrats have had both to develop and adapt to innovative structures and processes designed to provide politicians, academics and businessmen access to the levers of economic decision-making. Instead of embracing standard "off-the-shelf" democratic institutions such as parties and pressure groups, which it was felt would assert demands "detrimental to the national interest and development", the military and technocratic elite alliance created mechanisms such as the Council of Economic Ministers and a variety of national policy committees which were designed to provide wider access to policy-making while still depoliticizing the system and maintaining the technocratic quality of the decision-making structure. "The consultation between state and non-state agencies in Thailand, therefore, revolves not around participant political institutions such as political parties and parliament but around a unique kind of arrangement which can be called political gradualism and the technocratization of economic policy-making." In this way, the trappings of a classic administrative state have given way to what Chai-Anan describes as a "liberal-technocratic polity".

Chung-Si Ahn sees indications of parallel movements affecting the "Korea Inc." model. In this case, the society-wide push for a more democratic political system is combining with the new found strength of private sector enterprises and unions and the pressures on the government—arising from a more open economy—to limit the scope of state-controlled economic activity and to change the way economic decisions are made. "The process of democratization has made it more difficult for government to intervene arbitrarily in economic activities of the private enterprises. At the same time, economic policies are now subjected to the overhaul of the opposition-dominated legislature, which has become much more powerful under the new constitution." Unlike the situation in Thailand, these forces in South Korea have resulted in the participation of more groups in the policy-making process; more open, public criticism of the government's plans; the frequent politicization of economic issues; and the opening of rifts between the political elite and the economic technocrats, with the latter having to lobby for support for their "rational" solutions and the former becoming increasingly wary about the short-term political cost of technocratic recommendations. While such developments do represent a significant break with the traditional economic decision-making model, Ahn sees this as a transitionary period and cautions the reader not to exaggerate the degree of the changes to date. "The economic policies of the current regime maintain a basic continuity with the state-centric model. Therefore, the pre-eminent role of the state and the central importance of the economic bureaucracy, such as the EPB

(Economic Planning Board) are a continuing phenomenon in economic decision-making."

In China, the substantial economic reform movement which has gripped the system since 1978 has focused primarily on the decentralization of economic decision-making power. This decentralization, according to Cao Yuan-zheng, has taken two forms: the devolution of power downward through the existing administrative structure to lower levels of government, and even to the enterprise level; and the development of new or reconstructed decision-making organizations (e.g., individual, collective and joint enterprises). These decentralization initiatives have had the effect of moving the system of direct control of the economy towards a system of indirect control. This movement has a number of noteworthy features, as Cao points out. First, despite the fact the state as a whole has become less prominent in the process of making economic decisions, where it does remain an active player, the role of the government has become increasingly more important than that of the Communist Party. Second, neither the party nor the government intervene as a matter of course in the daily decision-making of enterprises. Third, the National People's Congress, the courts and even pressure groups have become part of the economic decision-making system. Finally, as economic reform has deepened, there has been a trend towards the separation of economic and social decision-making. In spite of such developments, the uneasy co-existence of the "two tracks" of reform make it very difficult to judge the eventual impact of these reforms on what is at root an extreme form of the administrative state policy-making model. For example, it has been reported recently in the media that the diminution of the influence of the government and the party in the management of economic enterprises is being reversed with the re-emergence of the importance of the government plan and the re-introduction of the party cadres in decision-making processes even at the enterprise level.

Other so-called administrative states represented in the following pages have been more cautious—apparent reform or propaganda to the contrary—in the modification of their bureaucrat-dominated policy-making structures. The long standing bureaucratic polity in Hong Kong, caught between the "upsurge of pressure politics" on the part of its own anxious citizens and the scheduled return of sovereignty to China in 1997, has become, according to H.C. Kuan, a "lame duck" administration, adverse to any meaningful structural reform or policy initiative. "There are also signs that the bureaucrats are demoralized and that departments are choosing to refrain from suggesting policy directions. The danger that looms on the horizon is that the government may lose the will or the ability to govern."

There is similarly little sign of structural innovation in Malaysia. Mavis Puthucheary argues that the "planned development approach has, as its dominant feature, the setting of substantive social and economic goals. This means that the main decision-making function is carried out within the government. Although the government is influenced by pressure groups and political claimants, it is the source of all policy innovations in the system." Furthermore, continuing adherence to the principles of the National Economic Policy (NEP) has had the effect of narrowing the decision-making group within government, excluding or significantly reducing the influence of non-Malay political leaders. The involvement of the Malay political and administrative elite in key public and private sector enterprises further enhances its power in the policy-making system. While it is true that recent privatization initiatives, coupled with the fostering by the government of the "Malaysia Inc." concept, have opened up the policy-making process to some extent, Puthucheary concludes that economic policies in Malaysia "are likely to continue to be formulated by the same small group that make up the political and administrative elite."

Singapore remains equally fixed in time, retaining many of the classic attributes of an administrative state. Linda Low's paper provides a detailed account of the evolution of this highly centralized and bureaucratized economic planning and policy-making system, tracking the rearrangements and movements of various key agencies and individuals, and the desultory efforts in recent years to increase consultation with the private sector and academics in the context of new deregulation and privatization initiatives. However, the pace of privatization is slow and the interventionary powers of the Cabinet and its bureacratic advisors remain substantial. As Low concludes, "State agencies and influence still predominate despite the delinking and privatization attempts and larger involvement of private sector initiatives."

This same generalization seems likely to apply to Indonesia, at least until the departure of Soeharto. Sjahrir provides an overview in his paper of the powerful role played by technocrats in the key economic ministries and the "centrality of the government in the national economy" through three different policy periods since 1967. He devotes significant attention to the most recent period (post-1983) which has been dominated by the struggle over deregulation and privatization. In this period, the policy battle lines have been drawn between the pro-deregulation forces (a coalition of some government technocrats, free market academics, business interests disadvantaged by regulation, and the major international financial and aid institutions) and those opposed (including some government technocrats, key leaders in the ruling political party, and those many members of the business community who continue to benefit from the distorted markets resulting from regulation and state enterprise). Sjahrir concludes: "Evaluating the impact of the ten deregulation packages achieved so far, it would not be wrong to state that the regulation forces have been weakened somewhat. However, their voice in the political power struggle appears to be steady . . ." Regardless of the eventual outcome of this struggle, the message from Sjahrir's paper is that the prevailing pattern of government influence over the Indonesian economy and the economic decision-making processes is unlikely to be substantially modified in the near future.

## Common Themes and Shared Shortcomings

It is obviously dangerous to try to find points of similarity or evidence of collective trends in the evolution of the economic policy-making systems of 12 distinct societies joined for the most part by little more than the wispy tentacles of a global economy and the artifice of membership in the "Asia-Pacific" region. Nevertheless, one cannot read the chapters which follow without being struck by certain interesting points.

First, as noted at the outset of this introduction, the forces at work on the governmental structures of virtually all of these jurisdictions are powerful. It is a rare state indeed that is not seeing its economic decision-making system buffetted by either by the neo-conservative "revolution" (in the form of downsizing, privatization and deregulation) or increasing public demand for more participation in policy-making, or both simultaneously.

Second, one has to be impressed by the continued diversity across these twelve jurisdictions of the efforts to improve policy-making processes. There are few cases in the pages which follow of blind allegiance to one model or another. Indeed, in the face of the kinds of heretical challenges confronting governance systems in Eastern Europe and across the globe, flexibility and pragmatism would seem to be essential characteristics of those who manage our public sector decision-making structures.

Third, in an era dogged if not dominated by neo-conservatism, the role of state economic planning has faced an onslaught of criticism. Virtually every one of the papers in this collection makes reference to planning; very few of them are complimentary. In a number of cases, there have been obvious reductions of the influence of the plan and the planner. In other jurisdictions, the planning system continues to operate, but the plans which are produced lose their imperative quality—at least with respect to the establishment of economic policy. There is a strong—but not universal—note of despair about the technocrat, unloved by everyone except the international aid and financial institutions and reduced (like any other player in the policy-making game) to lobbying their increasingly suspicious political masters.

Fourth, decentralization of economic decision-making power is the major theme of the paper on China, and an important subtext in many others. Decentralization takes many forms. Administrative decentralization involves the downward movement of authority and responsibility within one bureaucracy. Geographic decentralization is a variation on this, entailing the regional dispersal of decision-making authority. Organizational decentralization implies the devolution of power to satellite agencies of the government or to lower levels of government. In an important sense, privatization and deregulation are forms of decentralization as both usually involve the transfer of public sector decision-making power to private individuals or organizations. The chapters which follow are filled with expressions of anxiety related to decentralization in its many guises. For instance, will a government that decentralizes economic decision-making power have the capacity to operate effectively in a fast moving global economy? How do you implement policy without access to powerful publicly-controlled enterprises or regulatory agencies? Will privatizing just mean that you will have to regulate more? Is the private sector in a state-centric society capable of taking on many of the tasks now handled by the state?

Finally, the uncertainty expressed in several papers about the usefulness of structurally linking social and economic policy-making can be seen as a sign of a deeper malaise. For example, with a single exception, these papers are devoid of concern about the emerging issue of integrating environmental and economic decision-making. Embedded in the rhetoric of sustainable development is an institutional dilemma which will eventually require a rethinking of the relationship between the government and the economy, and a radical redesign of the machinery of government decision-making.

# Australia

## Australian Economic Policy: Problems and Processes*

*Michael Keating and Geoff Dixon*

### Setting the Scene

During the 1980s there have been major changes in economic policy in Australia. In particular:

- for the first time since 1960, real average earnings declined over an extended period of time, falling by 1.9 percent in both 1985-86 and 1986-87 and by 0.6 percent in 1987-88;

- for the first time in the 35 years for which comparable records have been kept, the Commonwealth has recorded a budget surplus; thus a deficit of $8.0 billion (4.2 percent of GDP) in 1983-84 was transformed into an estimated surplus of $5.5 billion (1.7 percent of GDP) in 1988-89, with reduced outlays rather than increased taxes bearing the brunt of restraint; and

---

\* The original version of this paper presented to the IRPP workshop in December 1988 was commissioned by the ACT Division of the Royal Australian Institute of Public Administration (RAIPA). The full text has been published separately as a monograph in Australia, the main difference being a considerable expansion of the discussion of micro-economic reforms. The final version of both the present chapter and the monograph have benefited from comments at a RAIPA seminar and the IRPP workshop and colleagues too numerous to mention here. Naturally final responsibility rests with the authors and their interpretation is not necessarily shared by their employer nor those who have provided comments.

- as perhaps the most dramatic of a series of structural reforms involving deregulation, the exchange rate was floated in 1983, and by the end of 1988 the Australian dollar had depreciated by 24 percent on the level prevailing from the beginning of the decade.

These developments occurred against a backdrop of greatly improved economic growth when compared with the 1970s and an associated strong rise in employment. Reflecting strong job growth, lower real wages were accompanied by rising real household disposable income.

This departure from Australia's historical norm of positive real wage growth and negative budget outcomes could hardly have been anticipated by an observer of the Australian economy during the 1970s. During that decade growth in wages (measured on an annual basis) invariably matched or exceeded increases in the Consumer Price Index (CPI), while changes in the Commonwealth budget outcome were confined to adjustments in the size of the deficit, with "belt tightening" exercises serving only to moderate the rate of growth in budget outlays.

What caused this remarkable turnaround in key policy settings, and what does it indicate about the nature and flexibility of economic policy formulation in Australia?

In addressing these questions this paper examines economic policy formulation in Australia over the last six years against the background of the theories of Olson and the new corporatists. The distinguishing feature of this period of Labour government has been an emphasis on consultation and "consensus", and in particular a prices and incomes policy based on an Accord between the government and the trade unions. This Accord has, however, evolved perceptibly in response to the major economic challenges of the 1980s. Also significant, but possibly less widely recognized, has been the extent of deregulation in a number of key areas of the economy.

The factors contributing to the pace of change in economic policy settings and to the greater emphasis on consensus-based processes are examined in the context of some of the more prominent policy issues of the 1980s. The conclusion is drawn that the concept of consensus based policies is less straightforward than may appear at first sight, and that neither the theories of Olson nor the corporatists fully explain the major changes in policy settings which have occurred.

Rather these changes have involved a unique blend of deregulation, aimed at improving the operation of markets, and "new" corporatism, aimed at more closely embracing business and unions in the policy forming process. In regard to the latter, an important factor contributing to recent major changes in policy settings has been the increasingly broad based (or "encompassing" to use Olson's term) nature of the key negotiating groups representing employers and employees. The impact of narrower groups, the interests of which may be more damaging to national economic goals, has been correspondingly curbed. Closely associated with these developments has been a process of learning from past failures, and an opening-up of the economy to international influences which has contributed to a more broadly-based appreciation of the nature of policy problems and the limits to what government can achieve on its own.

## The Australian Economy in Broader Perspective

Australia is one of the few countries to achieve a relatively high standard of living on a commodity export base. Although a member of the OECD it does not, in some respects, meet the criteria of an advanced industrial country and is not classified as such by the IMF. Unlike smaller advanced industrial economies such as Sweden, Holland and Austria, it has a vast raw material base. Commodities still account for two-thirds of exports and the economy is highly exposed to changes in world commodity prices. In foreign exchange markets the Australian dollar is widely regarded as a "commodity currency".

In contrast to most countries with comparably high living standards, the share of manufacturing in GDP (around 17 percent) is low compared to an unweighted OECD average of 23 percent in 1986. Reflecting this, Australian imports are dominated by manufactures, accounting for at least 60 percent of total imports in 1987. In the past, significant parts of Australia's own manufacturing sector have been accorded relatively high levels of tariff and quota protection, and this is now seen as having slowed the response of the Australian manufacturing sector to changing international markets. As in other high income countries, employment is dominated by the service sector, which accounts for around 60 percent of the total.

Reflecting the dominance of exports by primary products and imports by manufactures, the level of activity in the Australian economy is highly vulnerable to movements in the terms of trade, which have been declining in the long run. Economic growth, particularly development of Australia's rich resource base, has been reliant on long-term capital inflow, and this has produced a structural deficit in the current account. Moreover, resource development has tended to proceed in concentrated bursts, imparting an element of instability to Australian economic growth. During the 1980s a significant decline in domestic savings has meant an even greater reliance on foreign capital without an associated increase in the share of domestic investment relative to GDP.

While Australia's long-term reliance on foreign savings is more typical of a developing country, the economy has at the same time recorded unimpressive rates of GDP growth per capita compared with other OECD countries. An estimate by Maddison suggests that GDP per capita had fallen from 1.73 times the U.S. level in 1870 to only 80 percent of the U.S. level by 1982, with a growth rate at the bottom of 16 industrial countries over this period (Caves and Krause, 1984, p. 5). More recent evidence ranks Australian growth in per capita real GDP at 19th out of 24 OECD economies over the period 1960 to 1985 (Dowrick and Nguyen, 1987, p. 16) and since the mid-1970s productivity growth in Australia has been broadly in line with the OECD average (EPAC Paper No. 30, 1988, p. 12).

This historically unimpressive performance is variously ascribed to:

- high rates of population growth, absorbing capital for housing and social infrastructure and contributing to a spillover of employment into industries in which Australia has less comparative advantage;
- inability to take advantage of economies of scale due to Australia's small size and past tariff policy;
- unlike other countries, inability to benefit from the shift of resources out of (low productivity) agriculture;

- low productivity of capital investment, possibly reflecting the impact of the "tyranny of distance" on requirements for transport and communications infrastructure, excessive investment in housing and wasteful investment by governments;
- low investment in research and development and skills training; and
- the difficulty of achieving rapid growth in GDP per capita on a base which (reflecting early above average growth) was already high (Dowrick and Nguyen, 1987 and OECD, 1972, p. 23-28).

Slow growth in per capita GDP has not, however, been an aspect of Australia's performance which has triggered an explicit policy response. This is perhaps due to the high rates of growth which have been achieved in aggregate GDP (albeit in tandem with rapid population growth) and initially very high levels of GDP per head.

Inflation has also been a problem during the post-Second World War period, particularly since the early 1970s, when the first oil price shock and a short-lived commodity price boom were accompanied by a major wage explosion. While variations over time in the inflation rate have tended to reflect movements in international prices, particularly following the two oil shocks, Australia has generally experienced higher inflation rates than other advanced countries. Difficulties in restraining wage growth reflect high levels of unionization in the Australian labour market and the historical strength of the principle of "comparative wage justice", under which wage increases earned by stronger groups have been passed on to other groups in the labour market relatively rapidly. This in part reflects what Caves and Krause see as a strong egalitarian tradition in Australia (Caves and Krause, 1984, p. 2).

Australia's parliamentary system ensures that governments have considerable, but not in all respects certain, capacity to enact legislation necessary to implement economic policy. The government commands a majority in the House of Representatives (the lower house). It can therefore expect most of its legislative program to be passed. However, governments rarely have a majority in the Senate and are therefore not always able to secure legislation giving effect to particular aspects of policy. Failure to secure passage of the budget would lead to the downfall of the government.

In developing its policies the government receives economic advice from a number of departments of State (staffed by career public servants), and the Reserve Bank of Australia. Cabinet ministers also employ their own professional advisers. Several departments have separate research bureaux, and the increasing capacity of the bureaucracy for economic analysis has itself made a contribution to the coherence of policy formulation. The government also receives economic advice from a number of statutory bodies such as the Economic Planning Advisory Council (EPAC), the Commonwealth Grants Commission, the Industries Assistance Commission and the Inter-State Commission.[1] (A glossary of Australian economic policy-making institutions is provided in an Appendix.)

Organized groups can have a significant influence on economic policy in Australia through processes ranging from formal consultation to informal lobbying. The largest of these groups numerically is the Australian Council of Trade Unions (ACTU), which is an effective peak trade union body. Employer representation remains more fragmented, with three broad-based groups, the Confederation of Australian Industry (CAI), the Business Council of Australia (BCA) and the Australian Chamber of Manufactures (ACM) and smaller groups representing particular industry groups such as farmers, the metal trades and small business. Welfare groups, consumers and professional bodies also play a role where their interests are affected.

It is arguable that the influence of these organized groups has been enhanced in Australia by the frequency of elections. Federal parliament has a maximum term of three years, but in the last 16 years Federal elections have been called on average every two years. In addition, State elections can affect the political strategy of the national government and consequently its decision-making processes.

The environment in which Australian economic policy is made is influenced by a number of other features of the Australian Constitution. Most important, in the context of the current federal government's reliance on a prices and incomes policy, is the absence of any direct power of control over wages or prices. An Arbitration Commission, which is independent of the government, can determine minimum rates of pay for individual occupations, but employers can and do pay "over the award". Consequently, the government's prices and incomes Accord has had to depend upon voluntary adherence, without even the threat of compulsion, which may well have affected both the process of consultation and the nature of the trade-offs achieved.

## Features of a Federal System

Reflecting Australia's federal structure, State governments have a major influence on economic activity. At a macro level, the high degree of vertical fiscal imbalance which characterizes the Australian federation has been a fertile source of conflict. The national government raises around 80 percent of government revenues but is responsible for only 50 percent of final outlays, while State and local governments are dependent upon the Commonwealth for a significant level of assistance in order to perform their functions. This separation of the responsibilities for taxing and spending has arguably reduced the States' incentive for expenditure restraint.

While the Commonwealth has considerably reduced payments to the States along with restraint in its own expenditure in recent years, this has been reflected in a cutback in spending by the States of a capital rather than recurrent nature, with possible adverse long-term consequences for economic growth. Moreover, in recent years the division of responsibility and the increasing difficulty in reaching agreement with the States on restriction in borrowings by their semi-government authorities have affected the national government's ability to reduce the public sector borrowing requirement. All public borrowings are coordinated through the Australian Loan Council, and partly in response to a shift in focus away from the Commonwealth Budget deficit and towards the total public sector borrowing requirement, the Loan Council has exerted tightened control in the last few years. However, the two most recent Premiers' Conference/Loan Council meetings have been characterized by an unusually high degree of acrimony, culminating in the refusal of the State of Queensland at the 1988 Loan Council meeting to accept the borrowing allocation proposed by the Commonwealth.

To the extent that the federal nature of the Australian political system contributes to a shift in the burden of restraint onto the Commonwealth rather than other levels of government, distortions are likely to emerge in the distribution of restraint across spending programs at different levels of government. As mentioned, the balance of restraint at the State level between capital and recurrent spending is of particular concern.

The States can also significantly affect the supply side efficiency of the economy through their responsibility for business regulation, the activities of their substantial business undertakings, and the extent to which their investment in infrastructure encourages soundly based development. Again there are problems due to the overlapping of responsibilities between Federal and State levels of government. It is true that there has been a trend toward increased centralization of power with the Commonwealth, with specific purpose grants being used to achieve the Commonwealth's policy objectives through the agency of the States. However, the federal division of responsibility has led to problems of policy co-ordination in such diverse areas as transport and communications, company and securities regulation, resources development and marketing, energy policy, urban affairs, environmental policy, aboriginal affairs and taxation arrangements (Constitutional Commission, 1988, pp.63-73). For example, in the case of transport and communications there has been:

- proliferation of port facilities (particularly in South Australia and Tasmania);
- lack of commercial rail services; and
- poor road maintenance.

(EPAC Paper No 33, 1988, p.31).

It has recently been suggested that the Grain Handling Authority in New South Wales has annual export capacity far in excess of current tonnages, partly reflecting inadequate investment appraisal, while pricing policies have been inappropriately based on break-even rules and cross subsidization (New South Wales Commission of Audit, 1988, Appendix A4). The inadequacy of public authority pricing and investment policies is a recurring theme in the report of the New South Wales Commission of Audit, a matter of particular concern given the quite significant contribution of State semi-government authorities to GDP. Moreover, the principle of fiscal equalization between States under which the per capita level of Commonwealth grants to each State attempts to offset the geographical disabilities suffered by the State itself encourages non-optimal location decisions across the Australian continent. In Gramlich's words, "Overall, the Australian system of fiscal federalism receives high marks for equality, although not for efficiency" (Caves and Krause, 1984, p.273).

## Macro-Economic Policy
### *The First Phase of Restraint: Cuts in Inflation*

Australia, along with other developed economies, experienced a spell of slow economic growth in the 1970s, but performance *relative* to other economies also deteriorated during this decade. Over the ten years leading up to 1979, Australia recorded average annual increases in consumer prices of 9.8 percent (compared with 8.3 percent in the OECD) and in real GDP of 2.4 percent (compared with 3.3 percent).

This poor performance reflected problems encountered in the first half of the 1970s, particularly of an inflationary nature, together with the macro-economic strategy of "fighting inflation first" adopted in the second half of the decade (a policy common to a number of other countries experiencing the inflationary aftermath of the first oil shock). However, perhaps in recognition of market rigidities and their interaction with political realities, the avowed aim of the strategy was only a gradual unwinding of inflation. It involved moderately conservative settings (not always realized) for fiscal policy, control of the money supply through conditional projections of monetary aggregates (again frequently exceeded), and the government arguing for (but unable to engineer *directly*) falls in real wages.

This "inflation first" strategy was only partially successful. Some progress was made in reducing the rate of price increase, down from 15 percent in 1975 to 8 percent in 1978. However, reflecting the tight stance of fiscal and monetary policy, unemployment drifted upwards during the second half of the 1970s. Moreover the policy settings helped sustain the exchange rate at higher levels than might otherwise have occurred and insufficient progress was made in reversing the deterioration in competitiveness which had occurred in the mid-1970s. The failure to make greater reductions in inflation and unemployment simultaneously reflected a legacy of wage increases in 1974 which far outran productivity growth and which the "inflation first" strategy was unable to correct.[2]

In 1981-82 another major wages break-out occurred, reflecting a tightening in sections of the labour market and an expectation of an international resources boom which did not eventuate. At this time the Arbitration Commission played a small role in the determination of wages—typically it only ratified agreements *after* they had been reached by employers and unions, and often it played no role at all. The 1981-82 wage break-out was followed (as in a number of other world economies) by a sharp slide into recession in 1982-83 and further substantial increases in unemployment (full-time unemployment up by 47.6 percent in 1982-83).

As a result of these developments, by 1982-83 Australia's main economic indicators told a sorry story. Australia experienced negative real GDP growth of 2 percent in that year, an annual inflation rate of 11.5 percent and an unemployment rate of 9 percent. While the historically poor growth in income per head had been accepted as a cost of rapid employment and population growth, the rising unemployment which flowed from Australia's poor inflation and GDP performance in the 1970s, followed by the rapid deterioration in the early 1980s, demanded an effective policy response. This led the Fraser Liberal-National Party government of the day to commence the process of reversing the "inflation first" strategy, initiating an easing of fiscal policy and direct intervention in the labour market in the form of a negotiated wages pause. So began the process of change in economic policy settings which forms the subject of this paper.

This change in policy setting was taken up and extended by the Hawke Labour government elected in March 1983. Six weeks later the new approach was launched by a National Economic Summit Conference.

The concept of an Economic Summit represented a major innovation in Australian economic policy-making. It cut across the centralist and authoritarian stance of previous Australian governments and the consequently more adversarial nature of policy formulation which had developed, especially in the second half of the 1970s. Instead the Summit brought together the three major organized groups in the Australian community (government, employers and labour) along with the States, in an attempt to identify common ground in *simultaneously* reducing inflation and unemployment. The stagflation of the previous decade along with the bad experience of 1981-82, a new government, and the apparent reversal of policies by their predecessors meant that the Summit was called in most propitious circumstances. There was considerable sentiment in favour of any new approach.

Nevertheless when viewed against the widely differing approaches of employers and trade unions to resolving the economic problems of inflation and slow economic growth over the preceding decade, the outcome of the Summit manifested a remarkable measure of consensus. It was agreed by the major organized groups that the problems of high inflation and rising unemployment should be tackled *simultaneously* by a policy mix comprising:

- centralized wage fixation (following the breakdown during the 1981-82 wage explosion);
- a boost to business profits, with real wage maintenance pursued only as an objective over time; and
- fiscal stimulus to initiate a period of growth, and an effective prices and incomes policy to ensure that growth was not curbed by excessive price increases.

The Summit communique embodied significant changes to the positions which both unions and employers had adopted during the second half of the 1970s. In particular it embraced union readiness to accept a greater measure of wage restraint and improved profits, and employer willingness to accept centralized wage negotiation and a greater measure of fiscal expansion.

Reflecting both the Summit commitment to fiscal stimulus and the effect of cyclical recession, the general government deficit rose from 0.6 percent of GDP in 1981-82 to 4.2 percent in 1983-84. The Commonwealth budget deficit increased from 0.3 percent of GDP in 1981-82 to 4.2 percent in 1983-84, and the net PSBR rose from 3.4 percent of GDP to 6.7 percent. While it is possible to argue that the new government's fiscal policy only consolidated the stimulus initiated by their predecessors—the 1983-84 budget deficit was less than the $9.6 billion deficit allegedly inherited—the fact that the Summit ratified such a stimulus was especially important to a Labour government needing to establish its fiscal policy credentials.

The arrangement relating to wages was formalized in an Accord initially with the Labour Party and then between the government and unions. The Accord reflected a consensus based trade-off in which the ACTU agreed to wage restraint (including an immediate fall in real wage costs) in return for a number of social welfare measures, in particular the introduction of the new health care arrangements, tax cuts and tax restructuring. These changes effectively allowed for some rise in the social wage, even though money wage growth might be constrained to a rate which in the short term was below that of prices.

The shift to a consensus based strategy, involving negotiated solutions for fighting inflation and creation by the Labour government of a range of tripartite consultative mechanisms. These included the Economic Planning Advisory Council (EPAC), the Advisory Council on Prices and Incomes (ACPI), the Australian Manufacturing Council and its associated industry councils. These new bodies were built on a long tradition of government research agencies and quasi-independent bodies charged with the analysis of key aspects of the Australian economy. These have included the Industries Assistance Commission (IAC—previously the Tariff Board), the Arbitration Commission, the Grants Commission and the Foreign Investment Review Board (brief details are provided in the Institutional Glossary in the Appendix). This system has contributed to greater transparency and understanding of costs associated with particular policy options, assisting the process of reform in circumstances in which particular groups in the community have an interest in preserving the status quo.

Although the final powers of decision and responsibility remain with the government, the effect of the tripartite policy development arrangements introduced by the Labour government was to extend the mechanisms for scrutiny of public policy to macro-economic policy itself (an approach first recommended in the Vernon Report of 1965), and to institutionalize consultation with organized groups in the formal macro policy-making process.

In addition, a range of more informal contacts between senior government ministers and senior unionists and employers, particularly before major economic statements, were probably at least as important in forming and shaping a consensus based macro-economic policy.

At a minimum these various formal and informal contacts provided considerable opportunity for the government to explain its policies and seek support. Indeed, to achieve successful outcomes, consensus based arrangements do depend upon considerable government leadership; however, if the social partners are to collaborate fully, the government must also be prepared to listen— an appropriate balance of leadership and flexibility by the government is therefore required.

Almost immediately the consensus based approach was associated with an improvement in economic performance. Increases in real wages were curtailed while employment grew rapidly. Real GDP growth averaged 6.1 percent in 1984 and 1985 compared with an OECD average of 4 percent. The average annual growth rate of 6.1 percent over these two years was well above the 2 percent achieved in the previous six years (or 3 percent over the four years to 1981 if the recession years are excluded).

There was some debate in the mid-1980s about whether the successes of the Economic Summit and the Accord reflected a shift to a more corporatist approach to government, with business and employee groups collaborating in a "distributional coalition" (to use Olson's term) in which community based objectives override the narrow interests of individual participants.[3] Indeed, Schott has argued that countries adopting corporatist type arrangements were more successful in controlling stagflation during the 1970s (Schott, 1984, ch.4). Olson suggests that interest groups which encompass (for example) the great bulk of employees have a strong incentive to behave in a more socially responsible fashion than those which represent much narrower groups, since a greater part of the benefits of a socially responsible pattern of behaviour are "internalized" within the broader groups (Olson, 1982, p.48).

> Clearly the encompassing organization, if it has rational leadership, will care about the excess burden arising from distributional policies favorable to its members and will out of sheer self-interest strive to make the excess burden as small as possible. (Olson, 1982, p.48)

In this context it could be argued that the ACTU has in fact become a "more encompassing" interest group. From representing around 50 percent of unionists in 1949 it represented 85 percent in 1982.[4] Moreover, the ACTU had greatly improved its policy analysis skills and strengthened its ability to handle more extreme union elements in the early 1980s. The close involvement of the ACTU in policy-making under the new Labour government further strengthened its influence over its members.[5]

It might also be argued that employer groups have become more encompassing over the last decade. The formation of the CAI in 1977 brought together a wide range of employer groups representing the majority of employers, particularly in context of national wage case hearings. The BCA, representing the interests of the 50 largest Australian companies, was formed in 1983, partly in response to tensions between employer groups revealed in the context of the government's tripartite approach to policy formulation (McEachern, 1986, p.20).

The shift to more participative arrangements, and the enhanced authority of encompassing groups relative to particular sectional interests which they embrace, reflected a considerable disenchantment with the more separatist and consequently adversarial approach to policy formulation which had previously been adopted. The experience of a decade of stagflation together with the extremely adverse developments in regard to both inflation and unemployment in the period 1981 to 1983 are likely to have modified the perceptions of both labour and capital about the viability of policies of the type followed in the 1970s. Indeed, the first break with the conservative fiscal and monetary policies associated with "fighting inflation first" was made by the Fraser government in its last year in office in response to economic deterioration, while that government also called a tripartite meeting with unions and employers in July 1982 (at the request of the ACTU) to address the deteriorating economic situation. In Corden's terms, there was a learning experience for all concerned (albeit from somewhat different perspectives) about the adverse terms of the trade-off between inflation and unemployment achievable unless there was some adjustment to wage bargaining (Corden, 1988).

## The Second Phase of Restraint: Curbing the Growth in International Indebtedness

Despite the success of consensus based policies in improving the inflation-unemployment trade off and promoting a period of significant economic recovery in 1983-84 and 1984-85, Australian economic policy-makers were confronted by a further major challenge in 1985-86. This arose from a serious collapse in the terms of trade.

Australia has traditionally been a net capital importing country, with the current account deficit averaging around 2.5 percent of GDP in the two decades to 1980. Over this period there have been occasions when the current account deficit widened substantially, and as noted in the introduction, *both* inflation and the balance of payments have acted as constraints on growth. However, such episodes were generally followed fairly quickly by a closing of the deficit as the terms of trade and/or net export performance improved (Chart 1).

While the inflationary constraints on faster economic growth were eased by the peculiarly Australian consensus-based policies adopted in 1983, little had occurred to make Australian industry more competitive. In addition, the effectiveness of the post-Summit strategy meant that Australia began to grow rapidly relative to the OECD average. Thus, while rapid domestic growth sucked in imports, the slow pace of international growth was reflected in slow growth in exports. Earlier investment in new export capacity in anticipation of a resources boom stood underutilized. In contrast to the previous investment boom in the early 1970s, when increased overseas borrowings were quickly followed by higher export growth and earnings, on this occasion the large current account deficit persisted as the expected increases in commodity prices and hence capacity utilization failed to materialize. Effectively, the slow rate of growth in the world economy further impacted on the Australian balance of payments through a sharp decline in Australia's terms of trade.

Reflecting the combined effect of large overseas borrowings for investment projects which failed to generate the expected returns, large public sector borrowings, and a lack of international competitiveness partly reflecting wage/price inflation, the current

## Chart 1
### Current Account Deficit as a Percentage of GDP
### for Financial Years 1948-49 to 1987-88

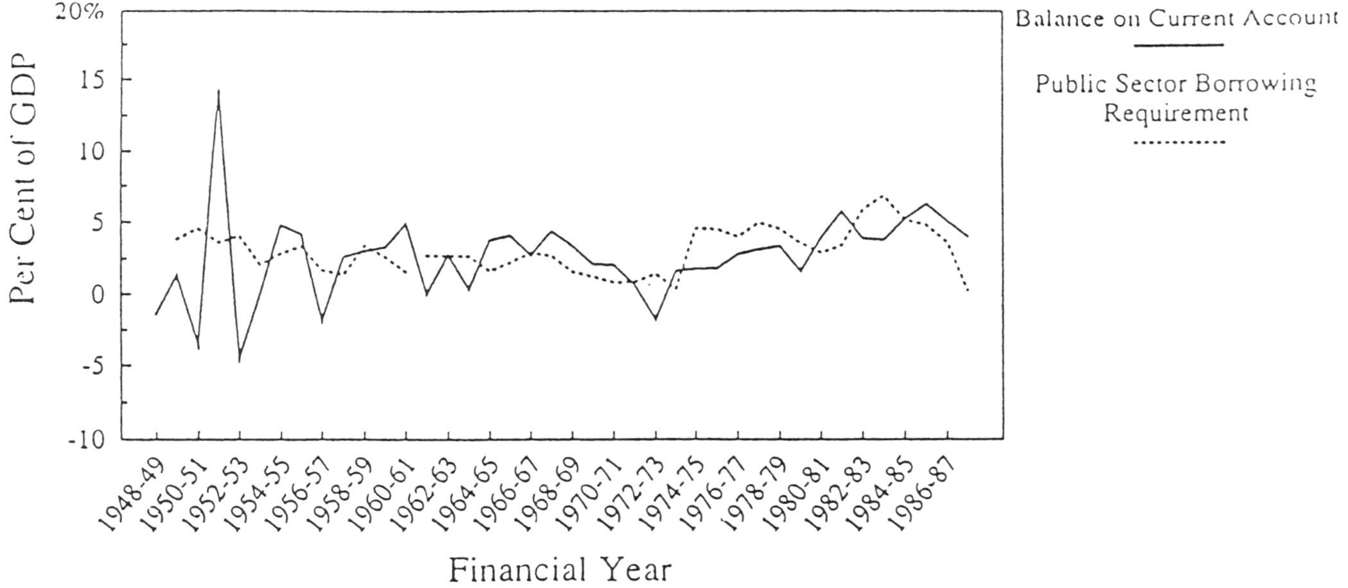

Balance on Current Account ———
Public Sector Borrowing Requirement ·········

Per Cent of GDP

Financial Year

Sources: Australian National Accounts ABS 5204.0 & 5206.0; Government Financial Estimates ABS 5501.0; Balance of Payments ABS Cat 5302.0 & 5303.0; Budget Papers, Various Issues.

account deficit averaged 4.8 percent of GDP over the six years to 1985-86. Structural rigidities caused partly by industrial policy also contributed significantly to this adverse current account performance.

In a sense, a formula for moving beyond the slow growth experience of the 1970s, based on restraint in labour costs, was beginning to emerge, but there was insufficient recognition of the other dimensions of competitiveness—especially damaging in a world in which slow growth remained the norm. Reflecting this, imports rose from 15.9 percent to 18.7 percent of GNE over the period 1983-84 to 1985-86 while exports as a percentage of GDP rose from 14.7 to only 16.1 percent. The resulting deterioration in the current account deficit was significant, from 3.8 percent of GDP in 1983-84 to 6.3 percent in 1985-86.

This sharp deterioration in Australia's trading position resulted in a rapid build-up in foreign indebtedness. Moreover, the deterioration precipitated a major depreciation in the Australian dollar, the TWI falling by a massive 38 percent between January 1985 and its low point in August 1986. This greatly increased the cost of debt servicing due to the denomination of debt in currencies other than the Australian dollar. Further significant adverse effects resulted from a shift in the composition of capital inflow from equity to debt capital.

By 1985 it was apparent that the post-Summit policy of Keynesian type fiscal expansion combined with a prices and income Accord was not a sufficient condition for sustainable growth.

The government's response to rapidly growing foreign indebtedness included two key changes to the policy settings previously agreed by the social partners at the Summit, aimed at stabilizing the growth in foreign debt relative to GDP:

- a call for a *decline* in real wages sufficient to "validate" the depreciation of the Australian dollar (necessary to prevent attempts to maintain real wages from neutralizing the competitive effects of the depreciation and from generating high inflation); and

- a reversal of the commitment to fiscal stimulus contained in the Summit communique. This involved engineering a decline in the public sector's claim on savings to an extent consistent with the requirements of foreign debt stabilization. It was to be achieved initially by firming up an existing "trilogy" of commitments involving, in particular, a reduction in the Commonwealth budget sector deficit.

The links between the public sector and current account deficits are complex. Chart 1 suggests a broad correlation between the two deficits over an extended period of time, although there is no clear annual relationship. However, given the identity of total saving and investment, it would be reasonable to expect that a decline in net overseas borrowing relative to GDP would necessitate a corresponding increase in public sector net lending, unless private sector net lending could be raised. Moreover, there was a requirement for strong growth in business fixed investment, particularly in tradeable activities, if the manufacturing sector was to be revitalized. To achieve this structural adjustment without drawing on overseas resources, and thus exacerbating external problems, there has been a need for further restraint on the public sector borrowing requirement. It is in this context that the government embarked on a further and more far reaching round of fiscal restraint beginning with the 1986-87 budget.

The process of restraining the outlays side of the Commonwealth budget has had the following characteristics:

- the government has endeavoured to ensure that the burden of adjustment in expenditure changes does not fall on the most disadvantaged in the community, involving increased targetting of welfare spending;
- where resources have been available for new programs, priority has been given to improving the social wage (increased family assistance, child care facilities and rental assistance) and assisting the process of industry adjustment, partly with a view to maintaining the support of the trade unions;
- where outlays have been reduced, emphasis has been given to generating further savings in the years beyond the year in which savings were first made;
- in a number of continuing programs or spheres of government activity, eligibility criteria have been reassessed to ensure that assistance is concentrated on those most in need (the assets test for pensions, the income test for family allowances);

- public administration has been streamlined and budget-dependent agencies have been required to offer up efficiency dividends; and
- resources have been devoted to reducing benefit abuse and fraud.

While the outlays side of the budget has borne the larger part of the burden of adjustment, revenue measures have also helped to close the gap. These have included:

- introduction of new imposts—capital gains tax and the fringe benefits tax;
- reduction of tax expenditures; and
- increased audit activity to improve the level of compliance.

The most important pre-conditions for fiscal restraint have involved establishing the political will and developing the necessary climate of public support. In regard to the first, ministers have been prepared to spend many hours thoroughly reviewing all programs in a Cabinet Expenditure Review Committee. A "May Statement" in advance of the August budget has been used to announce savings measures and facilitate their earlier implementation, while the public expectations associated with such a statement have assisted in establishing a savings target. In addition, expenditure restraint has been facilitated by the institution of a tight planning system based on publication of the next three years' forward estimates of outlays. These estimates are prepared by the Department of Finance and represent the minimum amounts *it* considers necessary to cover the costs of the government's ongoing policies and programs, with any changes to the estimates having to be publicly explained. This tight estimates discipline has meant that the previous experience of blow-outs in the estimates for future years has been avoided, with the estimates at the beginning of the last two budgets declining in real terms, substantially reducing the magnitude of the necessary savings task.

In order to build a climate of public support, the government capitalized on the drama of the rapidly depreciating exchange rate in 1986, which it attributed to the sudden fall in the terms of trade. This development was used to prepare the public for tough measures, with the Treasurer probably best capturing the public's attention when, in a celebrated remark, he warned that in the absence of such measures "Australia risked becoming a banana republic". Important support for the necessary adjustments and acceptance of their impact on income distribution was obtained from other opinion leaders, and most crucially from the ACTU. With the passage of time, however, the political opposition has been increasingly critical of the consequent reduction in some people's living standards and a policy of restraint has become more difficult to sustain.

By 1988 the government considered that the transformation achieved in the Commonwealth budget sector was broadly consistent with the requirements of foreign debt stabilization. Budget outlays have been reduced by 5.1 percent in real terms between 1985-86 and 1988-89, and from 29.8 percent of GDP to 25.6 percent. Over the same period revenue as a proportion of GDP ended virtually unchanged while the budget balance has swung from a deficit of $A5600m (2.4 percent of GDP) to an estimated surplus of $A5500m (1.7 percent of GDP). This must rank by world standards as one of the more remarkable medium-term fiscal adjustments in the post-war period, the more so because it was engineered by a party on the left side of the political spectrum.

## Linking Macro- and Micro-economic Policies

The events of the early 1980s—an upsurge in overseas borrowings to finance investment projects which subsequently did not meet original expectations, an expansionary fiscal stance, strong domestic demand by world standards and an unfavourable change in the terms of trade—all contributed to unsustainable growth in the current account deficit. However, a fundamental problem affecting the current account has been Australia's impaired international competitiveness since the early 1970s.

As Chart 2 shows, Australia suffered two major losses in competitiveness in the mid-1970s and the early 1980s as the nominal exchange rate rose and wage inflation accelerated. Each of these episodes seemed to be associated with major lifts in the import penetration ratio (Chart 3).

## Chart 2
## Australia's International Competitiveness

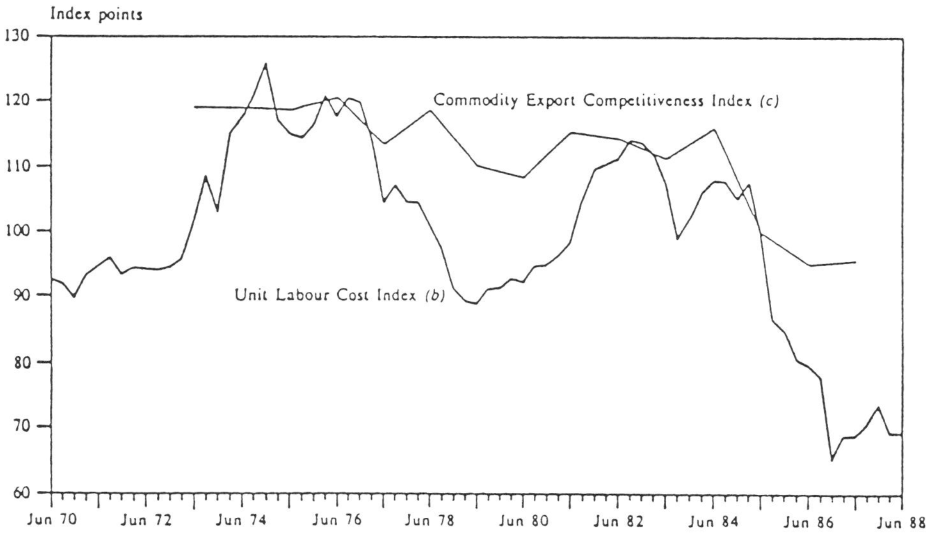

*Commodity Export Competitiveness Index (c)*

*Unit Labour Cost Index (b)*

(a) A rise (fall) in the indices implies a deterioration (improvement) in Australian costs and prices relative to our major trading partners and/or competitors after adjusting for exchange rate changes.
(b) The unit labour cost (ULC) index is the ratio of unit labour costs in the non-farm sector of the Australian economy to the weighted average of the exchange rate adjusted unit labour cost indices for the business sectors of Australia's four major import sources.
(c) The commodity export competitiveness (CEC) index is calculated using exchange rate adjusted changes in Australia's consumer price index relative to those for 28 other country and trading block competitors for 20 major Australian commodity exports.

Source: Budget Statements 1988-89, Budget Paper No. 1, p. 25.

## Chart 3
## Import Penetration Ratios: Domestic and Foreign

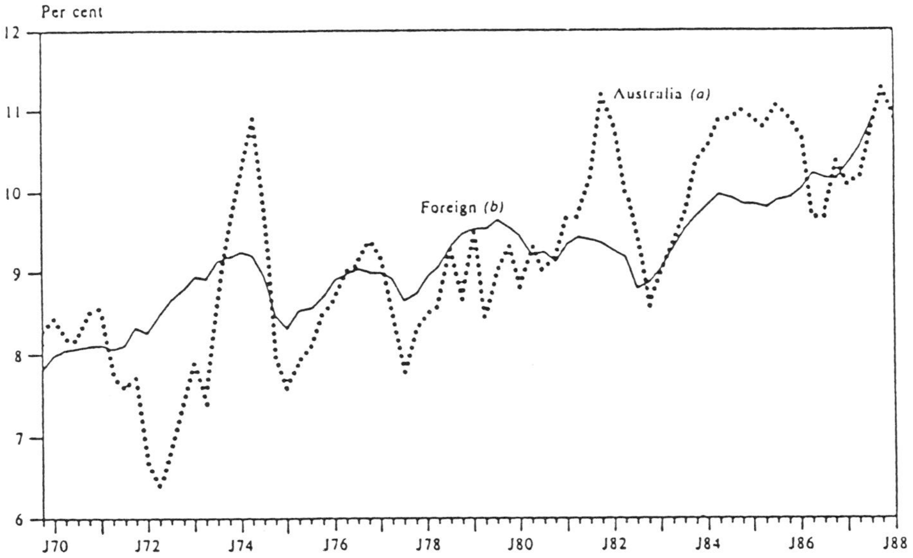

(a) Endogenous imports of goods as a percentage of non-farm sales, seasonally adjusted and at average 1984-85 prices.
(b) Ratio of goods and services imports to final sales. Indices for the United States, Japan, the Federal Republic of Germany, the United Kingdom, Italy and Canada are calculated on a 1984-85 base. An unweighted average of these indices is scaled by a constant factor to aid comparison with the Australian index.

Source: Budget Statements 1988-89, Budget Paper No. 1, p. 41.

Moreover, the competitiveness of Australian industry also suffers a legacy from past policies.

> In Australia's case, a framework of high protection and heavy regulation had built up over many years in major sectors. Inefficiency and inflexibility in factor and product markets reduced the economy's capacity to respond to external shocks and secular changes. An inward-looking industry perspective fostered inefficient business practices, excessive wage claims and a narrow export base. Heavy intervention in the operation of markets inhibited the movement of resources into areas that would yield the greatest benefits to the economy. The tax system was decaying and falling into disrepute. Insufficient effort was directed towards the development of labour and management skills. Superimposed on all that was a general reluctance to accept the need for radical change if Australia was to prosper in an increasingly integrated world. (Budget Statements, 1988-89, Budget Paper No.1, p.48).

Reflecting the above considerations, the share of manufacturing output in non-farm GDP has declined substantially from the level experienced during the second half of 1960s (Chart 4). This deterioration in the manufacturing sector has probably limited the country's ability to respond quickly to changes in economic circumstances.

An important recent development in Australian economic policy has therefore been the linking of the performance of macro-economic policy with the need for supportive micro-economic policies (rather than micro policies which may pull in the opposite direction in response to problems of a more sector-specific nature). This, in part, reflects a recognition that macro policy settings cannot achieve the desired combination of inflation, current account and unemployment outcomes regardless of the performance of markets at a more micro-economic level. The government's agenda for micro-economic reform has accordingly spanned the three main types of market activity: financial markets and taxation, product markets and labour markets.

The first area tackled was deregulation of financial markets, where the process of reform was in fact initiated by the former

## Chart 4
## Manufacturing Output Share and International Competitiveness

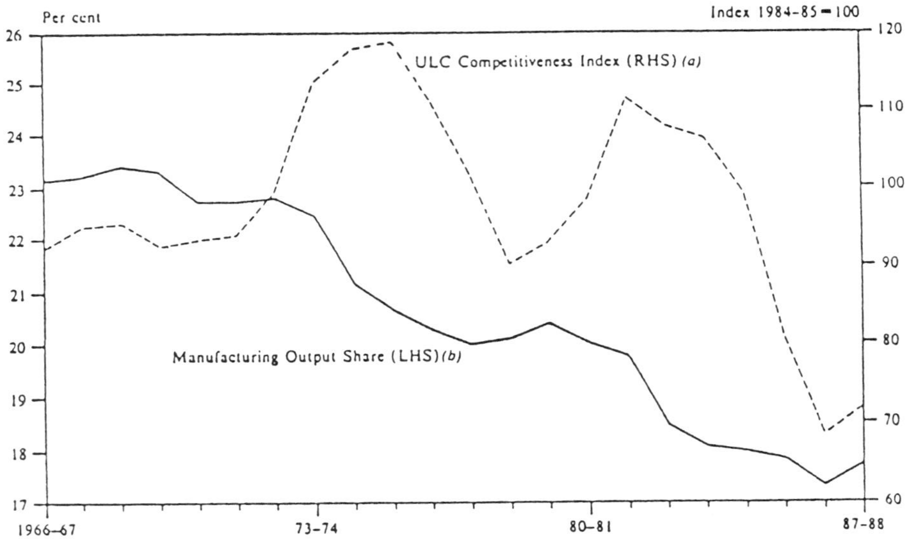

(a) The unit labour cost (ULC) index is the ratio of unit labour costs in the non-farm sector of the Australian economy to the weighted average of the exchange rate adjusted unit labour cost indices for the business sectors of Australia's four major import sources.
(b) Manufacturing output as a percentage of non-farm GDP at average 1984-85 prices. Figures for 1986-87 and 1987-88 based on Quarterly Indexes of Manufacturing Production (ABS Cat. No. 8219.0); the 1987-88 figure is based on three quarters data only.

Source: Budget Statements 1988-89, Budget Paper No. 1, p. 41.

Fraser Coalition government when it established a major inquiry into the Australian Financial System (The Campbell Report). Although the ground was thus well prepared, the catalyst for radical change may have been the pressure on the flexible band system of managing the exchange rate, to the point where the government was left with little choice other than to float the currency. Having taken this dramatic step, removal of controls such as those on lending rates, and expansion of the number of banking institutions, was less controversial. Moreover there was widespread recognition (partly due to the Campbell Report) that these controls were proving increasingly ineffective, as evidenced by the growth of new financial entities outside their scope.[6]

The major thrust of the reform of both taxation and industry policy has been to remove areas of special assistance and privilege in favour of establishing "a level playing field" on which all industries and enterprises can compete equally. Thus the thrust of taxation reform has been to remove particular concessions and in this way to broaden both the income tax and indirect tax bases. The acceptability of the reforms, however, has also been enhanced by the fact that they have not led to any net gain in revenue. This reflects a revamping of the income tax rate scale to reduce marginal tax rates impacting most heavily on higher income groups and an increase in tax and pension income test thresholds to protect the poor. In addition a key innovation introduction of an "imputation" system for corporate taxation which has terminated the double taxation of dividends.

Substantial differences in the incidence of tariff and quota protection between industries have also been an important feature of Australian industry policy in the past. Typically the commodity producing export industries have enjoyed little assistance, while the manufacturing industries have been protected, and for a few manufacturing industries the effective rate of protection has been more than 100 percent. The intention of the government's industry policy is now to produce a more outward-looking and export-oriented set of manufacturing industries, in contrast with the previous longstanding policy of import replacement behind protective barriers. The thrust of industry policy therefore has been a phased reduction in protection with special attention to highly protected industries such as motor vehicles and textile, clothing and footwear. Progress in such specially protected industries has partly been achieved by agreement to industry plans which involve a measure of government assistance for retraining, innovation and investment, along with commitments by employers to also invest and by trade unions to improve efficiency and work practices.

Because of Australia's vast distances, transport and communications plays an especially significant role in its economy. Moreover, there is good reason to believe that sections of the industry have imposed an undue cost on the economy. Much of the industry is publicly owned and/or involves provision of public infrastructure. Improvement in the efficiency of transport and communications therefore has depended importantly on improvements in the performance of publicly owned enterprises such as the railways, airlines, shipping, telecommunications and postal services. Reform of these enterprises has involved devolution to management of more responsibility and authority to manage by removal of many former controls, instead requiring the achievement of a target rate of return on capital together with improvements in productivity. Where the enterprises have social obligations to provide community services at less than cost these will in future be separately accounted for and will no longer mask assessment of the enterprise's efficiency and performance. In addition some steps have been taken to improve competition: for example by removal of the two-airline agreement to allow new entrants into domestic air services and genuine scheduling and price competition, and by removing some elements of Telecom's previous monopoly. Many of the "benefits" of the previously regulated monopolies had been captured by their management and staff and the new commercial environment is forcing the review of employment conditions such as manning levels and work practices.

This leads on to the issue of labour market reform, possibly the most important and most difficult element of the government's overall micro-economic agenda. In particular the government's reliance upon an incomes policy has required even-handed commitment to restraint by all employees which has inhibited changes in wage relativities. The OECD has suggested, however, that judged by turnover ratios or wage dispersion, or the speed at which aggregate wages react to market disequilibria, "it does not appear that Australian labour markets are particularly rigid" (OECD

1987-88, p.70). Instead the main problem in Australian labour markets is not so much flexibility between firms but the inflexibility of labour use within the firm, and associated with that, a lack of career opportunities and investment in training.

The main thrust of labour market reforms has therefore involved the review of work practices, a very substantial reduction in the number of job classifications to enable multi-skilling and better career opportunities, and increased and more effective investment in education and training. Much of the detailed negotiation of such issues as work practices, job classifications and training has occurred directly between employers and unions, with the government performing at most a facilitative role. The main inducement to union participation has been the possibility of wage increases, while for employers it has been the prospect of enhanced efficiency. Nevertheless, because of the overriding concern about competitiveness and aggregate labour costs, the overall cost of these changes has been controlled by centralized guidelines negotiated by the government with the peak organizations and approved by the Arbitration Commission. In addition, the government has provided substantially enhanced assistance on a means tested basis to encourage school leavers to obtain further training and education, and has also provided additional funds to increase the supply of education and training places, with priority for particular skills in short supply.

In seeking a consensus regarding the nature of its various micro-economic reforms, the government has usually relied upon committees of inquiry and/or discussion papers. These various reports have been important in increasing public understanding of the issues and persuading the public of the need for, and the merits of, reform. Frequently, however, the process of discussion has also involved elements of negotiation, with the eventual reform package often providing compensation or offsetting benefits to those most affected by particular changes. For example, dividend imputation was a major benefit to high income earners who were probably most affected by the broadening of the income tax base to include fringe benefits, capital gains, foreign tax credits and the non-deductibility of entertainment expenses, while special packages of retraining and relocation assistance have been offered to the employees most affected by reductions in industry assistance and protection.

However, probably the most important factor contributing to wide-ranging support for the government's micro-economic reforms is that each measure is seen as part of a coherent package and as such each has been accepted as part of the necessary response to the slide in competitiveness which occurred during the 1970s and early 1980s. Recognition of that slide was somewhat belated and in a sense was forced by the deterioration in the balance of payments in 1985, but the government then seized the opportunity to highlight the need for structural adjustment. The point has now been reached at which there is an acceptance by the unions that future wage increases should be based on the need to maintain the competitiveness of the export industries and that structural reform at the micro-economic level is required to improve Australia's competitiveness and ensure long-term employment growth (ACTU/TDC, 1987, ch.3).

While an externally imposed shock may have provided the catalyst, the continued acceptability of the new policy direction has depended upon macro-economic and micro-economic policies being seen as complementary. Thus a stronger macro-economic environment (that is more economic growth) can both result from successful micro-economic adjustment, and reduce community opposition to such micro adjustment (since any rise in unemployment which results is likely to be more acceptable in a strongly growing economy). Moreover, a failure of micro-economic policies to encourage the adjustments required by macro-economic developments would seriously compromise the effectiveness of macro-economic policy (Fraser, 1984, p.232).

This drawing together of macro and micro policies to form a more comprehensive economic strategy is of some significance in understanding how policies based on reductions in real wages and a move to a Commonwealth budget surplus could command a measure of bi-partisan support. The government's willingness to assist the process of structural adjustment is likely to have contributed significantly to an environment in which the unions could accede to reductions in real wages and the shift of the Commonwealth budget sector into surplus. While a consensus based process was involved it was essentially a *bargained* consensus, in which goals of some sections of the group were traded off in the pursuit of primary goals of the group as a whole. "Central to the sustainability of policy has been the steady and now substantial employment growth achieved" (Budget Statements 1988-89, Budget Paper No.1, p. 55). The creation of more than one million new jobs to the end of 1988, combined with improvements in the social wage, has meant that, despite reductions in real wages, real household disposable incomes have not declined and the disadvantaged and others most affected by adjustment have been protected.

## Conclusion

The Keynesian revolution and the economic prosperity of the 1950s and 1960s encouraged the view that governments could be held responsible for securing full employment. By contrast the poor economic performance of the 1970s, and in particular the deteriorating trade-off between inflation and unemployment, led to a gradual realization in the 1980s that most economies are not sufficiently flexible for governments to generate full employment solely on their own initiative. In these circumstances there were two possible strategies for a new and more effective approach to policy formulation:

- the first involves a more active role for the government in securing the co-operation of the social partners to achieve necessary changes. It involves a search for greater consensus and requires a significant element of negotiation in return for their support and greater transparency of policy development; and

- the second is based upon a less interventionist stance by the government, with efforts to improve the efficiency of the market instead. It involves a significant measure of deregulation and an increased awareness that where governments do intervene, policy goals are more effectively achieved by working through market forces rather than against them.

These two broad strategies for policy formulation are frequently presented as alternatives, but since 1983 the Australian government has relied upon a judicious blending of both. While it is perhaps better known for the emphasis it has put on the search for consensus, the government's hand has been strengthened by its initiatives to improve the efficiency of markets through deregulation. In particular, the floating of the Australian dollar has played a central role in changing the climate of economic debate in

Australia. The previous managed system based on an adjustable band necessarily had close regard to market forces. But as long as the authorities were seen to be directly setting the exchange rate on a daily basis, a degree of insularity was encouraged, together with a misguided sense that the government could fix anything which went wrong. The special significance of floating the exchange rate lies in the message, which it has forcefully conveyed, that Australians are living in a competitive world and that no government can generate full employment if real unit labour costs are uncompetitive.

Other important deregulatory actions which have helped to improve market efficiency include the reduction in tariff and quota protection, tax reforms to remove concessions to particular industries and ensure a more level playing field, and moves to put government business enterprises on a more commercial basis, thereby improving resource allocation and reducing the potential for featherbedding. In addition, industry itself has been encouraged to increase productivity by improving its work practices, pay systems and training. These changes have been secured with the co-operation of the peak organizations while they have reduced the power of specific constituent groups of those organizations.

Apart from the "stick" of market sanctions. a key factor contributing to support for consensus-based policies has been their perceived success. While there can be no absolute standard of such "success", a substantial majority of employers and trade unions appear to believe that the present approach to policy formulation has worked reasonably well, especially by comparison with past experience. In particular the combination of spectacular gains in employment with some progress towards debt stabilization has been most important in obtaining union support for fiscal and wage restraint, while the associated improvement in profitability has provided a "dividend" for employers. Moreover, consensus-based policies once in place appear to improve the trade-off between inflation and unemployment, allowing simultaneous progress on both fronts, which in turn has reinforced the basis for consensus. This result would be consistent with Schott's conclusion that countries adopting corporatist policies have better handled the problem of stagflation during the 1970s and early 1980s (Schott, 1984, Ch. 4).

Thus on the major issues, each of the principal parties has thought that it had more to gain by joining a consensus than by staying outside. What distinguishes this approach to policy formulation, however, is that joining such a consensus implies a "willingness to trade" in the sense that the parties are prepared to negotiate concessions from previously firmly held positions, even where this involves compromising the interests of some sections of the group in pursuit of primary group goals.

Clearly the scope for such a "bargained consensus" depends on the strength of the mandate held by the leadership of the encompassing group. Where primary group interests are acutely threatened by a policy of non-cooperation, that mandate may be strong. However, where threats are less extreme, or there are lags in recognizing threats, the "mandate to trade" accorded the group leadership by the constituency may be more limited, and the scope for bargained solutions reduced.[7] The qualified nature of the mandate held by employer leaders (reflected in the fragmentation of peak employer groups) may have contributed to the difficulty experienced by the government in eliciting employer support, and in a few instances, such as the 1985 attempt to introduce a broad-based consumption tax and the Industrial Relations Bill 1987, a negotiated consensus has not proved possible. Difficulties with the unions were also encountered in the context of privatization. However in each of these instances others have questioned whether the assessment made by the dissenting party was in fact in its own best interests. Clearly the leadership of each of the negotiating parties is very important in making that is assessment and in convincing their constituents. It is interesting to note that each of these examples where full consensus was not achieved was especially controversial politically.

While the main thrust of the government's policies have obtained a considerable measure of support, the question remains whether this process of bargained consensus has worked well enough. The Australian economy is still very exposed and vulnerable to any international shocks. By mid-1988 Australia's net external debt was as much as 30 percent of GDP and still rising. This of course was because the current account deficit continued to be unacceptably high and it would have further deteriorated in 1987-88 had it not been for the unexpected strength of commodity prices. In the longer run, arguably the best prospect for putting the Australian economy on a sounder footing involves improvement in the competitiveness of the economy, and in these circumstances the pace of micro-economic reform is also of some concern.

This leads to the main criticism of the open and consultative approach to policy formulation, which is that it inhibits the necessary flexibility of government responses. First, the process of consultation and the building of consensus takes time, leading to slowness in reaction to externally-imposed crises such as the sharp deterioration in the terms of trade in 1985. A second, and almost inversely related issue is that the government may be forced to anticipate developments, including the unpredictable, by committing itself to future policies which later prove inappropriate, in return for promises by the other parties. Third, the consensus approach may lead to an inadequate response if finding a consensus becomes the objective rather than securing the necessary adjustment. A variant on this theme is that the government may give away too much—for example by way of tax cuts—in order to secure that necessary bargain. In short, the consensus approach has been criticized as inhibiting policy flexibility, with the response to the fall in the terms of trade, for example, being criticized as being "too little too late".

It may be argued, however, that the allegation of slow response times does not fully take account of the uncertainty which inevitably attaches to economic policy formulation. In particular, there are lags in recognizing changes in the configuration of key macro-economic indicators and in identifying the magnitude and timing of policy impacts. The loss of policy flexibility associated with becoming over-committed too early is also a problem for all governments, whether or not they specifically seek a policy consensus. Indeed some very non-interventionist and independent governments have committed themselves to medium-term financial strategies and monetary targets in the hope of influencing the expectations and thus the behaviour of employers and trade unions.

Finally, it is arguable that many of the concessions which the Labour government has "bought" represent good value for money. By the beginning of 1989 the overall level of taxation in Australia had not fallen, so that it is probable the tax cuts involved in the government's wage/tax trade-off would have been required anyway, with the government effectively making a virtue of necessity. A more difficult judgement relates to instances where the government may have felt constrained from pressing the pace of reform in one area, such as transport and communications or privatization, in order to achieve acceptance for its highest priorities such as wages policy.

The alternative to bargained consensus would, as already indicated, need to concentrate even more upon efforts to improve the efficiency of markets by strengthening competitive market forces. This would require a reduction in the power of the principal producer monopolies and perhaps especially of the trade unions. Any such reduction in union power would probably involve substantial legislative intervention to over-ride the constitutions of what have traditionally been regarded as free associations and it would take time. Moreover reduced union power probably could not be achieved by legislative means alone, implying a need for much faster deregulation of product markets, reinforced by reduced demand for labour which, judging by the experience of the 1970s, might be protracted. This raises the question whether the alternative policy strategy to a consensus based approach could achieve successful adjustment without significantly more pain, at least for some major groups.

By contrast, the outstanding factor contributing to the success of bargained consensus under the present Australian government has been the achievement of substantial adjustment (especially by past standards) while sharing the burden of that adjustment so that no particular group is especially disadvantaged. This has meant, for example, that both employment and per capita disposable incomes have increased while real average earnings have fallen, so that the disadvantaged and the unemployed in particular have been protected.

Finally, in a democratic society it has to be acknowledged that governments are always likely to have regard to the political acceptability of change, whether or not they seek formal consultation with major interest groups. Certainly governments have a responsibility to provide a lead and inform public opinion regarding necessary change, but in this respect the consensus-based policies have been highly successful. Indeed it is difficult to think of any other country which has handled a major devaluation better in terms of securing as big a reduction in real wage costs as

Australia achieved over the following three years. On the other hand, only time will tell whether this attempt by the government to negotiate policy changes has produced sufficient adjustment at a sufficient pace.

Against this background it is of interest to set the significant changes in the formulation of Australian economic policy in the 1980s against Mancur Olson's suggestion that a major change in established (economic) policy settings requires a shock of some sort in order to break the strength of entrenched interest groups with a vested interest in the status quo. Olson emphasizes the links between economic policy outcomes and the structure and strength of the organized groups involved, with government seen as mediating between the interest groups. He argues that interest groups (other than those of the more encompassing variety) constitute a barrier to change, and he sees low growth in Australian incomes per head over time as reflecting the strength of organized labour and of the "protectionist lobby" in manufacturing. He proposes that an exogenous shock is required to break the strength of established interest groups before significant changes in policy settings can occur (Olson, 1982 and Olson, 1984).

It is true that there were major shocks to the Australian economy in the early 1980s in the form of the wage explosion and severe recession, the collapse in the terms of trade from 1985 and the major depreciation of the Australian dollar. However, rather than directly breaking the power of the groups which might have an interest in opposing change, these shocks were reflected in an augmented status for peak or more encompassing organized groups for which (in terms of Olson's analysis) there is less of a conflict between their own interest and broader social goals. The adoption of policies characterized by corporatist overtones served to reinforce the role and status of these more encompassing groups, contributing to an environment in which the pressure exerted by narrow sectional groups was more effectively countervailed. Given their broader base the peak groups had a "self-interest" in varying previously committed positions in response to Australia's declining macro-economic fortunes, and in embracing broader solutions than had been admitted during the 1970s. Changing circumstances served both to underline the perceptions within the peak groups of the limits to what was achievable under previous policies, and to increase their influence relative to narrower sections with a vested interest at variance with the group's primary interests and national economic priorities.

It is worth dwelling on the role of tripartism in this context. At least two benefits are likely to have been generated by the more tripartite orientation of government in the 1980s. Firstly, the closer involvement of the peak organized groups in economic policy formulation, which resulted from the more participative approach since the early 1980s, is likely to have reinforced the influence of those *more encompassing* groups for which national and group interest are more closely identified. In the ACTU's opinion:

> The sectional and parochial interests which in the past diverted attention away from developing a coherent and comprehensive set of national policies, have been obviated to some degree by initiatives of the present government. (ACTU/TDC, 1987, p.89)

In Olson's terms there is a much smaller "excess burden" on society when the encompassing group pursues policies in the interests of its members (Olson, 1982, p. 48). The corresponding reduction in influence of more narrowly based sections may, for example, have some bearing on the re-orientation in protection policy (which until the 1980s appeared to have been heavily influenced by the lobbying of the narrower sections directly affected). The capacity of the government to achieve significant changes in the settings of protection policy, such as have occurred since 1982, might therefore be better explained in terms of the growing influence of more encompassing interest groups than by the disappearance of narrower groups with a vested interest in protection.

Secondly, and also likely to have been of importance, was Corden's concept of a *learning experience* forced on the peak groups by the extreme events of the period 1981 to 1985 (Corden, 1988, p. 17). This learning experience related to a better perception of the limits to achievements possible without cooperation and in the face of incompatible (rather than irrational) expectations. Perceptions on the part of an organized group (including the government and its party) that its existing policy stance may no longer be optimal in a changing environment may take time to develop, particularly among the constituents of the group. Recognition lags and learning processes may be long and drawn out, postponing the possibility of consensus *between* different organized groups on the terms of a bargain for achieving radical changes in policy settings.

The speed of the learning process might therefore be an important factor in determining the stage at which agreement can be reached between different groups on major changes in policy settings. However, the incoming Labour government was helped by a considerable volume of literature which contributed to the learning process by making the costs and benefits of alternative policy options more transparent. The opening up of the Australian economy and society to international influences and opinion has also been very important. The significance of floating the Australian dollar has already been emphasized, but more generally, as Australia has become integrated into the world economy and Australians less insular, the Australian policy debate has become more influenced by overseas developments.

Not surprisingly this better appreciation of economic experience and literature has had its impact on the government's own bureaucracy. What is perhaps surprising is the extent to which, starting with the efforts of agencies such as the IAC and the Treasury, this bureaucracy took the lead in producing much of the literature and fostering the debate which has facilitated the learning process. This has also meant that among the government's "econocrat" advisers there was by the mid-1980s, a considerable measure of consensus and they seem to have achieved a greater influence than in many other countries. This represents quite a contrast with past debilitating debates within the government regarding the direction of such crucial policies as industry assistance and industrial relations, and which reduced the capacity for concerted action in the 1960s and early 1970s.

Thus by the 1980s there was a widespread recognition at least amongst opinion leaders that protection could lead to a net loss of jobs, that financial regulation and the tax system were not working as intended, and, most importantly, that wage increases did not necessarily lead to improved living standards but could cost jobs. Dissemination of this assessment by opinion leaders was then required to engender broadly based support within the community at large. This is the crucial role of political and industrial leadership and the quality of those leaders has been very important in this regard. They have had the responsibility for articulating the issues and appropriately structuring the debate to produce the desired outcomes. They have had to seize or even create the opportunities for achieving policy change. Much has depended upon their efforts to persuade the community about the need to adjust and on their ability to sell specific changes in policy to their constituents.

In sum, it has been the increased authority of encompassing groups and their capacity to adjust their position to protect primary group goals in the light of the turbulent events of this period, and to sell the changes to disadvantaged sections of their constituents, which provided the basis for the major redirections in policy. Moreover, an important benefit of a consensual (or tripartite) approach to government is that the inclusion of such groups in the policy development process obliges them (and their constituents) to focus on evidence relating to the relevance of their policies which may be ignored in a more adversarial environment.

Perhaps the truth therefore lies somewhere between the positions of the corporatists and Olson. A corporatist approach to policy development is unlikely to be a sufficient condition for realizing change where the options most likely to command broad support nonetheless impose significant costs on one or more major organized groups, and the learning process is insufficiently advanced for them to appreciate that "there is no (acceptable) alternative". On the other hand it is difficult to see narrowly based interest groups as *necessarily* constituting a barrier to change (as Olson suggests) in the light of the neutralizing effects on narrower groups of more encompassing interest groups which may emerge in time of crisis, and for which there is much less of a divergence between the goals of the group and those of society.

Part of the reason for Olson's scepticism about the scope for evolution in policy development is that he seems to believe that narrower groups are more likely to survive because they can provide greater benefits to their specific constituents—pursuing a greater share of the national cake will yield them a better return than cooperating to increase the size of that cake. The Australian experience does not on the whole support this pessimism and the encompassing groups have grown in stature and influence in a tripartist environment aimed at responding to major economic challenges. There have, however, been some attempted breakaways by a few ACTU constituents while business groups, notably the CAI, have remained fragmented to a degree, which emphasizes again the crucial role of the leadership in maintaining adherence within encompassing groups.

Recent Australian experience suggests that creating the institutions to support a "consensual" approach to policy development is of itself unlikely to manufacture common ground. It may, however, both increase the status and influence on policy outcomes of the more encompassing organized groups invited to participate (the interests of which are more closely aligned with the interests of society as a whole), while also helping to speed up the learning process within such groups on which scope for a bargained consensus may eventually depend.

## Notes

1. Unlike the other bodies the Inter-State Commission owes its existence to the Constitution (s.101).

2. A "real wage overhang" persisted through to the mid-1980s, perpetuated by a mixture of partial and full wage indexation until the early 1980s (Russell and Tease, 1988, p. 3).

3. Crouch sees the central characteristic of corporatism as a willingness of interest organizations to "constrain and discipline their own members for the sake of some presumed 'general' interest as well as (or even instead of) representing them" (Crouch, 1983, p. 452). However "authoritarian corporatism" is distinguished from "bargained corporatism" —in the latter case each group accepting "joint responsibility for the order and progress of the system as a whole" in exchange for concessions which further that particular group's interests (p. 457).

4. See Rawson, 1982, p. 104; the Australian Workers Union affiliated in 1967; the Australian Council of Salaried and Professional Associations disbanded in 1979, the Council of Australian Government Employee Organisations in 1981 and the Australian Public Service Federation in 1985.

5. The Prime Minister had been President of the ACTU from 1971 to 1980 and for more than a decade previously had been its industrial advocate. Another former ACTU advocate, Ralph Willis, was Minister for Employment and Industrial Relations (1983-87) and for Industrial Relations (1987-88).

6. By 1978 the regulated commercial banks held only 39 percent of the nation's financial assets compared with 52 percent in 1953.

7. Professor Parker has suggested in correspondence that the mandate held by the group leadership may also be reinforced by the recognition that bargaining strength is increased by the combination of those with allied interests, and by the reciprocal effect of greater integration within any one group on the cohesion of the groups with which it interacts.

## Bibliography

ACTU/TDC Mission to Western Europe. *Australia Reconstructed* (Canberra: AGPS, 1987).

Accord by the Labor Party and the Australian Council of Trade Unions Regarding Economic Policy (1983).

Australian Chamber of Manufactures, Industrial Relations Research Secretariat, (1987) *Bulletin No. 14, Corporatism, the Accord and the Employer Organisations in Australia* (Melbourne: Australian Chamber of Manufactures).

Australian Financial System Inquiry. *Final Report*, (Campbell Committee, 1981).

Budget Statements 1988-89, Budget Paper No. 1. (Canberra: AGPS).

Caves, R.E. and Krause, L.B. "Introduction" in *The Australian Economy: A View from the North*. (Sydney: Allen and Unwin, 1984).

Committee of Economic Enquiry. *Report on the Committee of Economic Enquiry* (Vernon Committee), 2 volumes. (Canberra, 1965).

Constitutional Commission. *Final Report of the Constitutional Commission: Summary*. (Canberra: AGPS, 1988).

Corden, W.M. *Australian Macroeconomic Policy Experience*, 1988 Australian Economics Congress Invited Paper, 1988.

Crouch, C. Pluralism and the New Corporatism: A Rejoinder." *Political Studies*, Volume 31, p. 452, 1983.

Dowrick, S. and Nguyen, D.T. "Australia's Postwar Economic Growth: Measurement and International Comparison." *Centre for Economic Policy Research, Australian National University, Discussion Paper* No. 160, 1987.

Economic Planning Advisory Council Paper No. 30. *Australia's Medium Term Growth Potential.* (Canberra: AGPS, 1988).

Economic Planning Advisory Council Paper No. 33. *Economic Infrastructure in Australia.* (Canberra: AGPS, 1988).

Fraser, B.W. "The Treasury—Tendering Economic Advice." *Canberra Bulletin of Public Administration*, volume XI(4), p. 230, 1984.

Gramlich, E.M. "A Fair Go: Fiscal Federalism Arrangements" in R.E. Caves and L.B. Krause (eds). *The Australian Economy: A View from the North.* (Sydney: Allen and Unwin, 1984).

Gruen, F.H. "How Bad is Australia's Economic Performance and Why?" *Economic Record*, volume 58, p. 162, 1986.

McEachern, D. "Corporatism and Business Responses to the Hawke Government." *Politics*, Volume 21, No. 1, May, pp. 19-27, 1986.

National Economic Summit Conference. "Documents and Proceedings," volume 1; "Government Documents," volume 2; "Record of Proceeds," volume 3. (Canberra: AGPS, 1983).

Nevile, J.W. *What's Wrong with the Australian Economy*, Centre for Economic Policy Research, paper presented at the 1988 Australian Economics Congress, Canberra.

New South Wales Commission of Audit. *Focus on Reform—Report on the State's Finances*, 1988.

Norton, W.E. and McDonald, R. "The Decline in Australia's Economic Performance in the 1970s: An Analysis of Annual Data." *Australian Economic Papers*, volume 22, 1983.

Organisation for Economic Cooperation and Development, *Economic Surveys of Australia*, 1972, 1985, 1987-88.

Olson, M. *Australia in the Perspective of the Rise and Decline of Nations.* Centre for Economic Policy Research, Discussion Paper No.109, 1984.

_____ *The Rise and Decline of Nations: Economic Growth, Stagflation and Social Rigidities*, (New Haven: Yale University Press, 1982).

Rattigan, G.A. *Industry Assistance; The Inside Story*. (Melbourne: Melbourne University Press, 1986).

Rawson, D. "The ACTU: Growth Yes Power No," in K. Cole (ed) *Power, Conflict and Control in Australian Trade Unions*. (Melbourne: Penguin Books Australia, 1982).

Russell, B. and Tease, W. *Employment, Output and Real Wages*, Reserve Bank of Australia Research Discussion Paper No. RDP 8806, 1988.

Schott, K. *Policy, Power and Order: The Persistence of Economic Problems in Capitalist States*. (New Haven: Yale University Press, 1984).

Sinclair, W.A. "The Australian Policy Tradition—Protection All Round," in J.A. Scutt (ed) *Poor Nation of the Pacific: Australia's Future*. (Sydney: Allen and Unwin, 1985).

## Appendix: Institutional Glossary

*The Australian Chamber of Manufactures (ACM)*

Established in 1877, the Australian Chamber of Manufactures (ACM) is the only national employer organization representing the manufacturing industry in Australia. As a result of its merger with the New South Wales Chamber of Manufactures on 1 July 1988, ACM operates a Victorian and New South Wales Division which jointly represent some 8500 members.

*The Australian Council of Trade Unions (ACTU)*

The peak trade union body, formed in 1927 to provide industrial advocacy services to affiliated unions. Representation at the ACTU congress increased from around 50 percent of unionists in 1949 to 85 percent in 1982 following the merging of peak councils for white and blue collar unions. Represents the union movement before the Arbitration Commission.

*The Business Council of Australia (BCA)*

Formed in 1983 by the amalgamation of the Business Round Table and the Australian Industries Development Association, the BCA comprises Australia's fifty largest companies. In contrast to the

CAI, companies are direct members of the BCA and are normally represented at the chief executive level.

### The Confederation of Australian Industry (CAI)
Established in 1977 following a merger of the Associated Chamber of Manufactures of Australia and the Australian Council of Employers' Federations, the CAI consists of three Councils:
- the Industrial Council, based in Melbourne, which is responsible for industrial relations, labour matters and related social issues;
- the Manufacturing Council, based in Canberra, which is responsible for developing and implementing policies concerning manufacturing industry; and
- the Commerce and Industry Council, based in Canberra, which is responsible for trade, industry and commerce policy issues not specifically related to manufacturing.

The CAI has represented industry on senior government advisory bodies, as well as representing private employers at hearings of national significance before the Conciliation and Arbitration Commission.

### The Arbitration Commission
Established by the Commonwealth in 1904 for the prevention and settlement of industrial disputes extending beyond the limits of any one State. Panels are allocated responsibility for an industry or group of industries. Cases of particular significance such as national wage claims, alteration of hours of work or leave provisions are dealt with by a Full Bench of the Commission.

### Commonwealth Grants Commission
An independent statutory authority established in 1933 and appointed by the Commonwealth to recommend upon the distribution of Commonwealth funds to the States.

### Department of Finance
Established in 1976 following the splitting of the Treasury, advises on and promotes the efficient and effective use of—and accounting for—resources to the Commonwealth government sector in accordance with government policies and priorities.

### Department of the Treasury
Advises and assists the Treasurer, and through him the government, in the discharge of their collective responsibilities for the management of the Australian economy.

*Economic Planning Advisory Council (EPAC)*

A council created in 1983 to draw together Commonwealth, State and local governments together with representatives of union, business and social groups for the purpose of discussing economic policy matters, and advising the government on these matters. The Council is chaired by the Prime Minister.

*Expenditure Review Committee (ERC)*

The Cabinet Committee which considers the budgetary implications of all proposals either to spend or save. Effectively the key Committee for formulating the government's economic strategy.

*Foreign Investment Review Board (FIRB)*

Established in 1976, the FIRB's primary function is to assist the Commonwealth government in administering foreign investment policy. The Board examines proposals by foreign interests to undertake direct investment in Australia and makes recommendations to the government on whether those proposals are consistent with the government's policy. It also provides guidance, where necessary, to foreign investors so that their proposals may conform with government policy. The FIRB is assisted by an Executive which is part of the Commonwealth Treasury.

The Board's functions are advisory only. Responsibility for administration of the government's foreign investment policy and for making decisions on proposals rests with the Treasurer.

*Government Business Enterprises*

The major Commonwealth government business enterprises are:
1. Australian Telecommunications Commission (Telecom)
2. Overseas Telecommunications Commission (OTC)
3. Australian Postal Commission (Australia Post)
4. Australian National Airlines Commission (TAA)
5. Australian National Railways Commission (AN)
6. Australian Shipping Commission (ANL)
7. Pipeline Authority (TPA)
8. Snowy Mountains Hydro-electric Authority (SMHEA)
9. Snowy Mountains Engineering Corporation (SMEC)
10. Quantas Airways Limited
11. Commonwealth Banking Corporation
12. Export Finance and Insurance Corporation (EFIC)
13. Housing Loans Insurance Corporation (HLIC)

14. Aussat Pty Limited
15. Australian Industry Development Corporation
16. Australian Capital Territory Electricity and Water (ACTEW)
17. Commonwealth serum Laboratories Commission (CSL)
18. Health Insurance Commission (HIC)

*Industries Assistance Commission (IAC; formerly Tariff Board)*

Comprising both full time and part-time Commissioners, its role is to encourage the development and growth of efficient Australian industries that are internationally competitive, export oriented and capable of operating over a long period of time with minimum levels of assistance. This is achieved by industry studies, usually following public hearings, which make transparent the cost of industry assistance.

*The Inter-State Commission*

The Inter-State Commission exists by virtues of sections 101-104 of the Commonwealth Constitution. It previously operated from 1913 to 1920, and was re-established in 1984. The Commission is responsible for investigating issues relating to interstate transport (referred to it by the Federal Minister for Transport) and means of improving the efficiency and equity of transport, and developing a balanced national transport strategy and infrastructure. It has Royal Commission-type investigatory and information-gathering powers.

*Joint Economic Forecasting Group (JEFG)*

The Joint Economic Forecasting Group brings together on a regular basis officers of the Treasury, the Department of Finance, the Department of Prime Minister and Cabinet, Australian Bureau of Statistics and the Reserve Bank of Australia to review economic conditions and prospects.

*Loan Council*

Representatives of Commonwealth and State governments meeting annually to determine Commonwealth and State borrowing programs (excluding defence and short term loans) and, under the "global limits", the aggregate borrowing program of larger semi-government authorities.

*Research Bureaux*

A number of Commonwealth government departments have economic research arms which undertake analytical work relevant to the functions of the department. These include:

- Australian Bureau of Agricultural and Resource Economics;
- Bureau of Industry Economics;
- Bureau of Transport and Communications Economics.

*Reserve Bank of Australia (RBA)*

Australia's central bank responsible for the formulation and implementation of monetary and banking policy. Policy of the Bank is determined by a Board of ten members, including the Governor of the Bank and the Secretary to the Commonwealth Treasury.

The RBA is principal bank to the Commonwealth government, some statutory bodies and State governments; it is banker to banks and certain other financial institutions; it prints and manages the note issue. As agent to the Commonwealth, the RBA distributes coin, manages the Commonwealth's domestic borrowing programs and conducts registries for Commonwealth government securities. The RBA oversees Australia's foreign exchange market, and holds and manages Australia's reserves of gold and foreign exchange.

The RBA's general responsibilities for the stability of the financial system provide the basis of its role in the prudential supervision of banks. The RBA operates in financial markets to influence the availability and cost of funds, and in the foreign exchange market where its purchases and sales of foreign currencies affect the exchange rate.

*Trade Practices Commission (TPC)*

The Trade Practices Commission is an independent authority, the function of which is to:
- advise business and consumers of their rights and responsibilities under the Trade Practices Act 1974 (as amended);
- to deal with possible breaches of the Act (compliance);
- to consider and determine applications by business for exemption from certain restrictive trade practices provisions of the Act.

# Canada

## Economic Policy-Making in Canada: The Case of the Canada-U.S. Free Trade Agreement

*A. R. Dobell*

### I. Introduction: The Economic Priorities of the Mulroney Administration

Traditionally, economic policy and economic policy formation in Canada have been discussed under the conventional headings of stabilization policy, resource allocation, and distribution. Under the first heading, underpinned by some version or other of an eclectic Keynesian macro-economic model, one examines the processes establishing the overall level of government expenditures and revenues, the resulting budget deficit, and its financing. Much has been written about this process of determining the appropriate levels of expenditures and revenues within some form of Planning, Programming, Budgeting System (PPBS) machinery, the resulting fiscal elbow room, the possibilities of expenditure management within some Cabinet "envelope" system, problems of consultation and secrecy within the budget process, and so on. Substantively, the "monetary-fiscal mix"—the problem of balancing the monetary policies of an independent central bank fearful of inflation against the fiscal stance of an administration apprehensive about the political costs of too-rapid reductions in budget deficits—is critical.

Under the second heading, the micro-economics of government policy-making is dominant. Structures of tariffs and subsidies, regulatory oversight and intervention, enterprise development grants and loans, licensing, and similar program initiatives are usually included in this discussion, with the major concerns, of course, being problems of regulatory "capture", "rent-seeking" behaviour and the distorting effects of such government programs on the decisions of individual economic agents. "Trade-distorting consequences" have to be balanced against other domestic policy goals, such as regional development or incentives for innovation.

In general terms, one can characterize much of the recent effort in economic policy in Canada as seeking to maintain an overall (macro-economic) balance in stabilization policy while opening up (micro-economic) programs with resource allocation consequences to market forces as much as possible.

In the third area—the link from economic policy to social goals—still more vexing and divisive issues arise. Under redistribution policy one addresses the structure of the tax system, the machinery for assuring security of incomes for those temporarily or permanently unable to work, for those retired, or otherwise disadvantaged, and similar questions of economic justice. Beyond taxes and transfer payments, the link to social programs may be extended to government service delivery activities like health and social services and daycare, as well as investment in education, family welfare, community infrastructure, and economic opportunity. In this area, of course, the concern has been with integration with labour market structure and alleged disincentives to saving, work and labour force participation.

In all this, it is generally accepted that the Department of Finance has responsibility for preparing the government's macro-economic strategy, fiscal framework and tax policies; the Treasury Board (a committee of Ministers) and its supporting secretariat, the responsibility for bringing together the spending plans of government departments and agencies into the expenditure estimates tabled in the House of Commons just before the start of each fiscal year (April 1); and the central bank (the Bank of Canada) for managing an independent monetary policy consistent with the government's overall fiscal stance and the needs of foreign exchange markets. One recent innovation was the adoption of a fixed target date (February 15) for tabling of the budget in the House of Commons by the Minister of Finance, so as to assure an orderly flow of public documents (Speech from the Throne, annual budget, annual expenditure estimates, and public accounts) setting out the formal plans of the government, and reporting on the realization of those goals so far as financial management is concerned. Individual departments, of course, manage pricing, regulatory, and subsidy questions relating to particular sectors.

Thus, the formation of economic policy in Canada evidently cannot be considered a simple process centred in any one agency. Nor can it be discussed in any comprehensive way in this paper. Instead, we shall narrow the discussion in two steps, focusing first on the broad priorities for economic policy as expressed by the present Progressive Conservative administration (the Mulroney government) when it first took office with a massive parliamentary majority in 1984. Then, we will narrow the scope still further, picking out, for illustrative purposes, what has proved the most dramatic (and traumatic) exercise in economic policy formation in Canada in the last half century—namely, negotiation of a free trade agreement with the United States. Examination of this process will illuminate a number of key features of contemporary economic policy-making in Canada, and suggest a few speculative elements for further discussion.

When the Mulroney government took office, three significant swings in economic and political thinking were evident in many nations. Among academic economists in the United States, a "supply-side revolution" had been born out of the decade of economic disarray from the late 1960s to the late 1970s, resulting in rejection of the well-established Keynesian view of macroeconomics, and adoption of a "new classical" view which denied that governments could exercise any constructive role through discretionary economic policy, and insisted on the efficiency of unimpeded market forces. One consequence of this swing was dramatically increased interest in reducing detailed government intervention in economic activity and opening up economic activities to market forces.

A second development, a direct legacy of the violent swings in inflation and real interest rates through the 1970s and early

1980s, was vastly heightened concern with problems of inflation relative to unemployment, and consequent focus on problems of public sector borrowing and deficit reduction. A third was a preoccupation with the alleged adverse consequences of high tax rates and generous income security measures for saving, investment, labour force participation, and labour productivity, as part of a general swing to a more conservative economic and political posture.

Thus, in November 1984, the new Minister of Finance released "A New Direction for Canada—An Agenda for Economic Renewal", focused on deficit reduction and control of inflation; decentralization, privatization, deregulation, and a general shrinking of the role of government; tax reform; and an opening of the Canadian economy to external market forces. The over-riding theme was to remove economic decision-making from parliamentarians and legislators, and restore it to the market place, where—in the view of the Minister and the government—it belonged. The key message was contained in the complaint that previous governments had "substituted the judgements of politicians and regulators for the judgements of those in the marketplace" (Canada, 1984, p. 2).

All of this can, of course, be seen as part of a renewed business agenda for economic policy, and part of a shifting balance of political power, reflecting a breakdown of the post-war consensus on social and economic policy which earlier had emphasized the goal of full employment. In place of full employment and income security, productivity and international competitiveness became the watchwords. Following the second oil shock and the engineered recession of the early 1980s, concern in Canada turned increasingly to the growing challenge—only dimly perceived initially—of the successful development efforts of Japan and the NIEs (newly industrialized economies).

One central initiative that emerged out of these concerns was the decision to negotiate a free trade agreement with the United States. We turn now to this focal event in Canadian economic policy-making of the last several years as an illustration around which to explore some of the contemporary problems of economic policy formation in Canada.

## II. Processes and Players: The Example of the Free Trade Agreement

The development of the Free Trade Agreement between Canada and the United States provides a good illustration of the informal and formal decision-making processes and the roles of a wide range of players in the Canadian economic-policy process. The purpose of what follows is not to provide a detailed account of developments leading up to the Free Trade Agreement, but simply to summarize those aspects of the story that illustrate some important elements of the economic-policy process in Canada.

As already noted, when the Conservative government took office in 1984, there was a sense among economists that trade liberalization should be a policy priority. Concern that the GATT system was weakening, that U.S. protectionism was increasing, and that there was a need to stimulate a more dynamic and competitive manufacturing sector in Canada led to this sense of urgency. Studies undertaken by various research institutes, including the Institute for Research on Public Policy (IRPP), the C.D. Howe Institute, and the Ontario Economic Council emphasized the need for more liberal trade policies in order to stimulate the manufacturing sector and to make it more competitive; that is, they emphasized the lowering of Canadian tariff walls to open the Canadian economy to international competition.

Business perspectives on free trade were central to this process. As a result of the earlier Tokyo Round of GATT negotiations, Canadian tariffs had been substantially eased. This, together with the highly successful Auto Pact, prompted changes in the manufacturing industries in Ontario and Quebec, who then had improved access to American markets. By the 1980s, these industries had successfully adjusted to reduced tariffs and found that they could specialize quite effectively in order to cope with tariff changes. As a result of their success, the manufacturing industry, and the Canadian Manufacturers' Association in particular, moved from an ambivalent stance on free trade to strong support for the trade negotiations with the United States. Other business lobbies were also very concerned about the threat of U.S. protectionism. Of particular note is the Business Council on National Issues (BCNI), which represents 150 of the largest corporations in Canada (many of them American multinationals), and has considerable influence with the government. From the outset of this process, the BCNI has been perhaps the strongest voice advocating the need for a comprehensive free trade agreement to assure access to U.S. markets.

In Canada, the use of Royal Commissions has been a popular way to bring government and external expertise together for exploration of policy issues in controversial areas of particular concern to the government. Royal Commissions operate independently, hold public hearings, and their findings are made public. Their recommendations occasionally figure prominently in the policy-making process. The Royal Commission on the Economic Union and Development Prospects for Canada (popularly known as the Macdonald Commission, after its Chairman) was mandated to examine contemporary political structures and economic development prospects for Canada and played a key role in the processes leading up to the Free Trade Agreement. The members of the Macdonald Commission, who were appointed in 1983 and began work early in 1984, made the decision to focus their report on a limited set of topics: the economy; federal-provincial relations; the structure of the economic process; and the constitution, recognizing, above all, the need to examine these issues in a global context. Many groups made submissions to the Macdonald Commission during their hearings, including labour groups, business lobbies, women's rights groups, research institutes, organizations of first nations, social lobbies and environmental groups.

The commissioners emphasized two dimensions of international trade for Canada—multilateral trade, and bilateral trade with the U.S.—and recognized, with respect to the former, that any new multilateral trade negotiations would be part of a very slow and uncertain process on which Canada could exert only marginal influence. The Commission concluded in part that:

> Canada's economic growth is critically dependent on secure access to foreign markets. Our most important market is the United States, which now takes up to three-quarters of our exports. More, better and more secure access to the U.S. market represents a basic requirement, while denial of that access, an ever-present threat. We are extremely vulnerable to any strengthening of U.S. protectionism. Early bilateral

negotiations with the U.S. could provide opportunities for the two countries to negotiate reduction or elimination of tariffs and other barriers to cross-border trade, at a pace and on a scale not likely to be achieved multilaterally in a further GATT round. Such negotiations could also be used to win agreement on rules designed to deal with special or unique problems affecting cross-border trade; they would provide a more secure shield against a U.S. policy of protection. (Macdonald Commission Report, Vol. III, p. 379).

There was only one dissenting voice among the commissioners on the free trade issue. This was Gerald Docquier, the leader of the steel workers' union, whose opposition reflected, for the most part, the views of the labour movement on this issue.

The Mulroney government did not come into the 1984 election in support of free trade; in fact, free trade was not an election issue and indeed, the new Prime Minister was on the record as strongly opposed to the concept. The government did, however, have an interest in improving relations with the U.S. They tried to follow through on efforts made by the previous government to develop free trade agreements with the U.S. in particular sectors, but were not successful. Not surprisingly, the sectors identified by American negotiators as candidates for sectoral free trade agreements were different from those which were of concern to Canadians.

Increasingly, some senior officials in key positions in the Department of External Affairs (which also carried responsibility for international trade matters), in concert with officials in the Department of Finance and the sectoral economic departments, began to conclude that a comprehensive Free Trade Agreement was the only viable alternative, particularly given growing U.S. protectionism and the prospects for growth of regional trading blocs around the world.

As the support of business, academic, bureaucratic and media players for the free trade idea strengthened, and with the legitimacy given to the concept of a comprehensive free trade agreement by the Macdonald Commission, the government began tentatively to test plans for some specific initiative with key business and other interests in Canada and with the U.S. This led, in September 1985, to the formal announcement of Canada's willingness to enter into comprehensive free trade negotiations with the U.S.

The government set up a separate organizational structure to deal with the task of negotiating the Free Trade Agreement. This structure included a new, special-purpose Trade Negotiations Office, which would undertake the actual negotiations; regular meetings of federal and provincial first ministers to discuss progress and provincial concerns about free trade; an internal committee of ministers to oversee negotiations; and a system of sectoral industry advisory committees. Both the sectoral arrangements and the federal-provincial relationship deserve comment.

A network of 13 sectoral groups was established to advise the negotiators on issues affecting their particular sectors. The focus groups in these and other consultations were the leading elements of the business community, as well as certain other opinion leaders. There was no effort made to exclude grass-roots organizations and other groups, but neither was any substantial effort made to create a vehicle to bring them into the process. The Canadian Labour Congress (CLC), which represents most of organized labour in Canada, boycotted the consultation process from the start. The CLC opposed the Free Trade Agreement and provided a large share of the funds and energy in the lobby against free trade, arguing that the Free Trade Agreement would induce a shift in power further favouring big business, and that Canadian jobs would be significantly threatened. (On the other hand, the much smaller Canadian Federation of Labour (CFL), which represents the older craft unions, argued in favour of the Free Trade Agreement.)

In the past, the federal government has assumed that its international trade policies would gain acquiescence from the provinces with minimal consultation. However, the debate over the Free Trade Agreement spurred the evolution of the provinces' role in international economic negotiations. Provincial governments in Canada play an increasingly important role in the formation of national economic policy, primarily because of the economic levers which are available to them. A milestone in federal provincial relations was the Federal-Provincial First Ministers' conference in Halifax, in November 1985. The principle of provincial participation in negotiating free trade was inaugurated at this meeting (although the international negotiations themselves continued to be carried out by the Federal negotiators). The free trade negotiations marked the first time that the provinces have been so deeply involved in the international economic negotiating process. Regular meetings of the First Ministers were held throughout the negotiating process to review progress and flag issues, and ongoing consultations were undertaken between federal and provincial officials to underpin this process. In the end, only two provincial premiers opposed the Free Trade Agreement that was finally negotiated (albeit one of those provinces—Ontario—is the largest and most economically powerful in Canada).

During the parliamentary process to put into legislation the agreement that had been reached between the two governments in October, 1987, both opposition parties attempted to block its passage, arguing that the deal which had been negotiated threatened Canadian sovereignty. But they were unsuccessful, and the government used its large majority to force the legislation through the House of Commons. Many complained that the agreement was pushed through Parliament much too quickly, and that it had not received adequate scrutiny. Those who felt insufficiently involved in the consultative process were particularly vocal on that point. John Turner, Leader of the Opposition, then used his party's majority in the Canadian Senate (an appointed body similar to the British House of Lords), to block final passage of the legislation, on the grounds that an election was required to allow the public to decide on an issue as important as free trade.

That election was held in November, 1988. After an unprecedentedly volatile and negative campaign, the government was returned with a majority (the first time for a Conservative government this century) and hence, in the Canadian parliamentary tradition, with a mandate to implement the free trade deal.

But that majority was substantially reduced from the previous standings, and was based on a clear electoral lead in only one of Canada's five regions—Quebec, where political and business support for free trade was monolithic. Nationally, more than 57 percent of the votes were cast for parties strongly opposing the trade deal, interpreting it as a giveaway of Canada's resources and policy autonomy in exchange for not-very-secure access to U.S. markets, and only limited protection against unilateral application of U.S. trade laws.

So both the motivation and the mandate remain unclear and hotly contested, and the best one can say about this episode as an exercise in policy formation is that it tested the practical limits of parliamentary democracy.

## III. Problems in the Processes of Economic Policy Formation

This brief account points to some fundamental weaknesses in contemporary economic-policy processes in Canada, and also illuminates how that process itself has begun to evolve. It does have to be said that all of this discussion is very much in flux. As this is written, the lessons to be drawn from the massive controversy over the Free Trade Agreement cannot yet fully be discerned. Policy-making in Canada may be in transition to a much greater degree even than we presently realize. But there is no doubt that some major developments are in train.

The issue, at its briefest, is the need to find ways to involve more effectively the much broader range of economic and political players who will be necessary participants in formulating and implementing economic policy in the future. The relatively narrow band of players who could expect to carry the day in earlier times—senior federal ministers and officials, leaders of key Canadian corporations, and a small number of other opinion leaders—will no longer suffice.

As was noted earlier, the process leading to the Free Trade Agreement did try to expand the range of players involved— principally by increasing the involvement of provincial First Ministers and officials, and through more systematic consultations with industry in the sectoral advisory groups. However, many other groups, who in the end would help to determine the fate of the Agreement, were not involved, or at least felt excluded. Those who felt outside the process included most of organized labour, environmental groups, farm organizations, a variety of public-interest social lobbies and native groups. One question, then, that arises from the experience with the Free Trade Agreement for the future of economic policy-making in Canada, is how a wider range of players can be more effectively involved, without the whole process becoming so complex that nothing can be done.

Another question, which perhaps helps point the way to an answer to the first, is how to generate a vision of where Canada should be going—as a context within which particular accommodations and agreements can be negotiated. The absence of a widely shared understanding of why Canada was pursuing the Free Trade Agreement and of the overall vision of a Canadian future to which it was intended to contribute, meant that important issues were not debated or resolved at the outset; they returned in disguised forms at the end of the exercise to help increase confusion, fear and doubt about what the Free Trade Agreement contained and what its implications would be. Stated somewhat differently, if we had a clearer understanding of where we were trying to go, it would have been much easier for Canadians to evaluate the extent to which we had got there—or at least had identified an appropriate road heading in the right direction.

A further important question illuminated by the Free Trade Agreement experience is the importance of the transparency of the policy process. If a wider range of players are to be more effectively involved, the process by which they are to be engaged must be clearer to all, and the ways in which particular outcomes are determined equally should be apparent to all. It is increasingly unrealistic to expect people to trust the outcomes of a process they cannot see or understand. Thus the problem of confidentiality in negotiations processes as well as in budget processes must be addressed head-on: is the present extensive cult of secrecy surrounding these and other economic policy-making processes really essential, or even consistent with the broader consultation and social learning processes likely to be inevitable in the future?

In contrast to these requirements of consultation, consensus-building, and transparency, the Free Trade Agreement followed no normal channels of policy formation, and, indeed, it became caught up in, and totally confounded by, an unprecedentedly divisive and bitter national election. Currently, the policy-making process in Canada is idiosyncratic and ad hoc; the course to be followed in policy development is never very clear—indeed even at the centre it is not clear that any Cabinet planning system is being consistently pursued at this stage.

In addition, rather than forging a wide range of interests and perspectives into a consensus view of where Canada should be going, the government's economic policy agenda over its first four years tended to parallel very closely that of the business community, primarily big business and multinational enterprises. The Federal Government in power between 1984 and 1988 was by no means alone in Canadian history in taking this tack, but times have changed, and this more narrowly-based approach is now less likely to suffice in Canada for the making of economic policy.

In this context, when implementation of the negotiated free trade deal did become an election issue, opposition coalesced, not around the question of tariff protection, but around non-tariff barriers and the consequences of continental integration for independent Canadian policy discretion in social welfare, environmental quality, sustainable development, and—more generally—sovereignty. The question which emerged was whether the undeniable and inevitable forces of globalization and integration in the world economy meant an equally inevitable convergence of national policies in all areas with significant links to economic performance and international competitiveness, and hence an unavoidable drift toward continental integration in North America. Canadians had evolved no shared answer to that dilemma. That this development had not been sufficiently anticipated stemmed, in part, from the fact that the Conservative economic policy agenda is premised on an intellectual framework that questions—indeed largely rejects—the possibility of any constructive exercise of discretionary economic policy by governments. The constraints imposed by the negotiated Free Trade Agreement on such policy discretion, therefore, were not seen by the government as a disadvantage. Indeed they (along with dispute resolution procedures) were in many ways seen as the rationale for acceptance of a deal, even though that deal did not in fact succeed in entrenching secure access for Canadian goods and services to U.S. markets, while it did appear to commit Canada more tightly as a supplier of energy resources and raw materials to the United States.

What the free trade episode underlines, and what should be emphasized here, is that economic policy must be seen as embedded in broader issues of political economy (particularly with respect to social policy, including related issues of investment in human resources, and to environmental policy, including related issues of investment in renewable ecological resources). The problem is that there is no place in our processes of economic policy-making where any integrated and coherent decisions, rooted in some broadly-based consultative processes, can be developed and made to stick in dealing with issues of that scope.

This basic social dilemma is compounded by the fact that economic policy processes in Canada are largely centred in national institutions, while the issues increasingly reflect international developments not easily integrated into, or comprehended by, existing national institutions; and, at the same time, the relevant instruments increasingly seem to fall within provincial jurisdiction or at the level of community development.

We find, in other words, a classic mismatch between the reach or domain of the institutional structures for economic policy-making and the extent or scope of the substantive problems to be addressed. The time frames and spatial scales appropriate to management of the renewable resource base fundamental to Canadian society are not consistent with the time frames and spatial scales of economic policy management, nor of global commodity markets. The institutions of economic policy-making do not bridge these disparate domains.

## IV. Reform Trends and Possibilities

This very cryptic commentary suggests a general version of a dilemma that is not confined to Canada. How are we to reconcile independent national goals in domains like social policy or sustainable development with the fact of increasing openness in an increasingly integrated, interdependent, globalized economy? More particularly, as it is increasingly recognized that international competitiveness is rooted fundamentally in better social and organizational relationships, greater investment and utilization of human resources, and better stewardship of renewable and exhaustible resources, it becomes crucial to establish credible institutions for economic policy-making which go beyond the conventional market rhetoric. The fundamental question becomes the nature of the social contract—our collective commitments as to the relationship of individual to community and to economic enterprise, as well as our obligations to future generations. We don't yet know how to develop any shared views on such principles, let alone how to implement them at local, national or supranational scales of activity.

In such a setting, the community's longer-term goals and principles find no reflection in resource allocation decisions or the conduct of economic activities. This is not a new problem, of course; but what has changed is the degree to which economic decisions and economic policy-making must be integrated within these larger frameworks.

There are a number of interesting reform proposals that have been suggested in the Canadian context, including several arising out of IRPP studies, to break through some of the identified barriers to widespread consultation and synthesis of diverse interests in the formation of economic policy. These fall into four general categories: new devices for consultation on policies of the Canadian federal government; mechanisms for achieving better integration of domestic social and economic policy; national proposals for the integration of economic policy with ecological concerns, leading to a strategy of sustainable development; and possibilities for new international policy-making institutions. Brief reference to some proposals in each of these categories is presented here.

Within the context of economic policy-making in the Canadian federal government, reforms have been recommended to enable Canada to face more effectively the challenge of the coming century. The government will be forced to decide on new policies to guide economic development, and such decisions require a consensus regarding the goals of the country and the plausible targets for the economy, as well as an equitable sharing of the costs and benefits associated with new policies. James Gillies maintains that responsibility for creating such a consensus rests with the federal government, and that in a democratic, pluralistic nation such as Canada, the best mechanism for achieving this consensus would be a joint committee of the House of Commons and the Senate, which would hear the views of the diverse groups affected by proposed policies (Gillies, 1986). Furthermore, Gillies suggests that current decision-making processes should be supplemented by permanent sector forums, which would bring together experts from various sectors of the economy to review economic policy and offer recommendations to the government.

Proposals for more effective integration of social and economic policy emerged from an informal IRPP workshop on income security reform held in 1987. Particular areas of concern noted were the development of a coherent strategy for social policy at both federal and provincial levels; the maintenance of federal standards across regional variations; and the development of a balance between conflicting interests. The mechanisms proposed to achieve agreement on these issues included active federal-provincial consultation, and the creation of national fora on social programs to facilitate the exchange and dissemination of social policy data (Seward/Iacobacci, 1987).

Further proposals directed towards the increased integration of social and economic policies arose from an IRPP study of the social policy process in Canada (Dobell/Mansbridge, 1986). The study reached the conclusion that the capacity of institutions to respond to changing circumstances can be increased, and social forces redirected towards a more comprehensive examination of social policy processes in Canada. In this context, regular social reporting was recommended, in order to "shift present public orientation from a strictly economic framework to a more comprehensive, socio-economic understanding" (Dobell/Mansbridge, p. 42). Ongoing multipartite consultative processes, allowing substantial participation of all groups concerned in policy formation were also proposed, to address issues of social programs and socio-economic policy.

Among domestic proposals to achieve increased integration of economic policy and ecological concerns are a number of suggestions for reforms in government institutions. In response to the report of the World Commission on Environment and Development (the Brundtland Commission), the government of Canada established a Task Force (the Task Force on Economy and Environment) to bring together representatives of industry, government and environmental associations, to work towards a system in which economic initiatives are judged ultimately by their environmental soundness. In order for environmental considerations to gain greater acceptance in economic circles, a terminology analogous to generally accepted accounting and economic concepts must be developed for discussion of environmental considerations and values (Gale, 1988). In addition, as environmental assessment is a necessary instrument for sustainable development, a restructuring of the Federal Environmental Assessment Review Office (FEARO) has been proposed, in order to provide the agency with the increased scope necessary for it to screen general policies rather than simply specific, government-initiated projects (Hazell, 1988). Public participation, the power of subpoena, and autonomy from other federal departments would be essential characteristics of the new agency.

A study initiated by the Royal Society of Canada notes the lack of any federal body designed to consider long-range issues, with authority to examine the long-term consequences whose identification is an essential component of planning for sustainable development. In this context a Parliamentary "Committee on the Long-run" has been proposed, to address intergenerational environmental issues (Braybrooke/Paquet, 1988).

As global changes increasingly call into question previously-held notions of international security, and the need for international cooperation to overcome transboundary problems becomes more evident, several proposals for international dispute resolution mechanisms and policy-making institutions have arisen. Among these, the Canada-U.S. Free Trade Agreement itself is notable for its incorporation of a mechanism which could well prove to be a useful model of multilateral dispute settlement, by providing a forum for consultation, as well as a medium for settlement negotiation. A permanent binational Commission, with a defined time-table for each stage of the settlement procedure, and the obligation to submit differences to binding arbitration by an objective tribunal should negotiations fail, characterize this framework (Smith/Stone, 1987).

The need for international institutions designed to take into account key resources of human, ecological and information capital—all of which are very difficult to measure, and fall outside current economic accounts—as well as the economy-environment linkages central to international economic coordination, is becoming increasingly apparent. The tripartite forum of the Pacific Economic Cooperation Conference (PECC), which brings together members of the academic, business and government communities in order to facilitate more harmonious economic linkages amongst Pacific countries, is one such institution. As such, it should be supported and strengthened, and mobilized to counter the fragmentation of the world economic system. In order for this effort to be successful, cultural and educational exchanges among countries in the region must accompany economic cooperation, and a stronger organizational framework, incorporating an international secretariat, as well as supporting networks of research institutes, will be necessary. Both greater resources and effort should be committed to the support of PECC as an example of an effective information-based, consensus-building, non-governmental forum for multipartite discussion at the international level.

A specific proposal for international cooperation arose from the World Conference on "The Changing Atmosphere", held in Toronto in June 1988. This proposed "World Atmosphere Fund" would be designed to raise revenue for research into and development of alternate energy sources, and assistance of developing countries, while simultaneously discouraging fossil fuel use. Revenue would be raised through a one percent tax on fossil fuels for small sources (such as private automobiles), together with the sale of internationally administered permits for large emissions. Initially, some permits might simply be assigned low-income countries, and the rest auctioned. Permits would be tradeable, and would also be augmented by credits gained through sound forest management. Since the June 1988 meeting there have been several further international conferences scheduled for discussion of similar schemes, which amount to the formation of economic policy through international negotiation. The significance of this trend can hardly be overstated.

## V. Conclusion

In summary, we can identify three distinct observations flowing from this examination of the Canada-U.S. Free Trade Agreement as an illustrative example of economic policy-making in Canada.

First, it exemplifies the way in which the business agenda and economic interests evolved, under the pressures of globalization and the increasing influences of multinational enterprises, from a general preference for protectionist policies, to advocacy of an outward-looking open regime which would dramatically limit the scope for government intervention and discretionary policy in the face of market forces.

At the same time, however, business interests recognized that they would have to contend with a much more pluralistic and open policy process, and geared up to ensure effective participation in that political bazaar.

Similarly, business enterprises have come to recognize that they cannot ignore or dismiss the growing number of increasingly effective interest groups representing, or intervening on behalf of, environmental causes, aboriginal claims, health and safety concerns, or other such issues. Responsibilities of corporations toward this wider array of stakeholders increasingly are accepted, though with widely varying degrees of conviction.

Thus, the second feature illustrated by the 1988 Canada-U.S. free trade deal is the necessity of extending the policy process across a much broader and more disparate array of participants in a much more open, less predictable setting. No evident mechanism exists in Canada to bring such a multi-party, multi-interest process to a head in any orderly set of "peak negotiations" leading to decisions which will "stick".

And finally, the third distinct feature illustrated by the Canadian debate over the free trade agreement with the United States as a case study in economic policy formation is the extent to which longer term economic policy has to be set within the broader context of social policy, and cannot be divorced from concerns for environmental matters. (This means, of course, that such policies inevitably must be shaped by cultural traditions and moral values as well as by market forces, and political institutions must be found to give expression to these influences.)

In the end, however, perhaps the major conclusion emerging from this discussion still is that the answers to any of these questions now generally transcend national boundaries and can only be found in international relationships and rules. Paradoxically, then, even as greater efforts are made to elaborate much more comprehensive policies within a much wider, noisier and more participatory framework, perhaps less and less discretion may remain for national governments to carve out any effective independent economic policy, no matter how decisively negotiated or broadly accepted domestically.

## Bibliography

Amuzegar, Jahangir. *Comparative Economics: National Priorities, Policies, and Performance.* (Cambridge, Massachusetts: Winthrop Publishers Inc., 1981).

Auld, D.A.L. and Miller, F.C. *Principles of Public Finance: A Canadian Text.* (Toronto: Methuen, 1977).

Byran, Ingrad A. *Economic Policies in Canada.* (Toronto: Butterworths, 1982).

Braybrooke, David and Paquet, Gilles. *Human Dimensions of Global Change: The Challenge to the Humanities and the Social Sciences.* (Royal Society of Canada, 1988).

Bryce, R.B. "The Essentials of Policy-Making." *Policy Options*, Vol.2, No.4, 1981.

Clark, Ian. *Recent Changes in the Cabinet Decision-Making System.* 1986.

Dobell, A.R. *Multilateralism, Regionalism and Bilateralism in Asia-Pacific Trade: A Canadian Perspective.* Insert to September/October 1988 IRPP Newsletter.

_____ "Pressing the Envelope." *Policy Options*. Vol. 2, No. 5, 1981.

_____ "The Administrative Adjustment to Economic Downturn: The Canadian Experience." 1986.

Dobell, A.R. and Mansbridge, S.H. *The Social Policy Process in Canada.* (Montreal: The Institute for Research on Public Policy, 1986).

Dobell, A.R. and Parson, Edward. "A World Atmosphere Fund." *Policy Options*. Vol.9, No.10, 1988.

Doern, G. Bruce and Phidd, Richard W. *Canadian Public Policy: Ideas, Structures, Process.* (Scarborough, Ontario: Nelson Canada, 1988).

Doern, G. Bruce, Maslove, Allan M. and Prince, Michael J. *Budgeting in Canada: Politics, Economics and Management.* (Ottawa: Carleton University Press, 1988).

French, Richard D. *How Ottawa Decides: Planning and Industrial Policy Making 1968-1984.* (Toronto: James Lorimer and Company Publishers, 1984).

Gale, Robert. "Eco(nomic)ological Policy." *Policy Options*. Vol.9, No.3, 1988.

Gillies, James. *Facing Reality: Consultation, Consensus and Making Economic Policy for the 21st Century*. (Montreal: The Institute for Research on Public Policy, 1986).

Government of Canada. *Regulatory Reform: Making it Work*. (Ottawa: Office of Privatization and Regulatory Reform, 1988).

_____ *Currents of Change: Final Report, Inquiry on Federal Water Policy*. (Ottawa: 1985).

_____ *A New Direction for Canada: An Agenda for Economic Renewal*. (Ottawa: Department of Finance, 1984).

_____ Securing Economic Renewal:
- Canada's Economic Prospects: Looking Ahead to the 1990s
- Budget Papers
- The Budget Speech
- The Fiscal Plan

(Ottawa: Department of Finance, 1988).

_____ *A Review of Canadian Trade Policy: A Background Document for the 1980s*. (Ottawa: Minister of Supply and Services, August 1983).

_____ *Canadian Trade Negotiations: Introduction, Selected Documents, Further Reading*. (Ottawa: Minister of Supply and Services, December 1985).

_____ *Canadian Trade Policy for the 1980s: A Discussion Paper*. (Ottawa: Minister of Supply and Services, 1983).

_____ *Sales Tax: Measures Relating to the Treatment of Marketing and Distribution Costs – Technical Notes*. (Ottawa: Department of Finance, 1988).

_____ *Federal Programs and Activities: A Descriptive Inventory*. (Ottawa: Federal-Provincial Relations Office, 1987).

Hazell, Stephen. "An Environmental Auditor." *Policy Options*. Vol.9, No.3, 1988.

Holling, C.S. "The Biophysical Connection." *Policy Options*. Vol.7, No.1, 1986.

Kent, Tom. "Real Tax Reform." *Policy Options.* Vol. 9, No.3, 1988.

MacNeill, Jim. "Sustainable Economic Development." In *Canada-Japan: Policy Issues for the Future,* edited by K. Lorne Brownsey. (Halifax: The Institute for Research on Public Policy, 1989).

Maslove, Allan, M. *Tax Reform in Canada.* (Halifax: The Institute for Research on Public Policy, 1989).

Milne, G.D. and Taylor, Lynne. *Federal Policy Process.* (Ottawa, 1985).

McQuaig, Linda. *Behind Closed Doors: How the Rich Won Control of Canada's Tax System.* (Markham Ontario: Penquin Books, 1987).

Milne, Glen. *Federal Policy Process.* (Ottawa: Napean Development Consultants, 1986).

OECD. *The Control and Management of Government Expenditure.* (Paris: OECD, 1987).

Phidd, Richard W. and Doern, G. Bruce. *The Politics and Management of Canadian Economic Policy.* (Toronto: Macmillan Company of Canada Ltd., 1978).

Rawson Academy of Aquatic Science. *The Environmental Impacts of Government Policy.* (1988).

Raynauld, Andre. *The Canadian Economic System.* (Toronto: Macmillan Company of Canada Ltd., 1967).

Royal Commission on the Economic Union and Development Prospects for Canada. *Case Studies in the Division of Powers.* (Toronto: University of Toronto Press, 1986).

_____ *The Politics of Economic Policy.* (Toronto: University of Toronto Press, 1986). (Especially Section 1)

_____ *Division of Powers and Public.* (Toronto: University of Toronto Press, 1986).

Savoie, Donald J. *Regional Economic Development: Canada's Search for Solutions.* (Toronto: University of Toronto Press, 1986).

Science Council of Canada. *Water 2020: Sustainable Use for Water in the 21st Century.* (Science Council of Canada Report 40, June 1988).

Seward, Shirley and Iacobacci, Mario. *Approaches to Income Security.* (Halifax: The Institute for Research on Public Policy, 1987).

Simeon, Richard. "Global Economy/Domestic Society: Competing Challenges to Governance in Canada." In *Canada-Japan: Policy Issues for the Future,* edited by K. Lorne Brownsey. (Halifax: The Institute for Research on Public Policy, 1989).

Smith, David C. (editor). *Economic Policy Advising in Canada: Essays in Honour of John Deutsch.* (Montreal: C.D. Howe Institute, 1981).

Smith, Murray G. and Stone, Frank (editors). *Assessing the Canada-U.S. Free Trade Agreement.* (Halifax: The Institute for Research on Public Policy, 1987).

Wilson, Tom. "Canada's Economic Prospects Over the Medium Term (With an Analysis of the Impact of Tax Reform and Free Trade)." In *Canada-Japan: Policy Issues for the Future,* edited by K. Lorne Brownsey. (Halifax: The Institute for Research on Public Policy, 1989).

Wilson, Honourable Michael H. *The Canadian Budgetary Process: Proposals for Improvement.* (Ottawa: Department of Finance, May 1985).

# New Zealand

## The New Zealand Experience*

*Gary Hawke*

## 1. The Historical Background

For 30 years after 1938, New Zealand was able to satisfy the income aspirations of most of its citizens while emphasizing the creation of a more diversified economy which could provide employment opportunities for people with a wide range of interests and aptitudes. Private investment was high. Business was confident that new plant and equipment would be justified by market expansion, more through population growth than through increased average per capita incomes. If necessary, import licensing would be used to protect the domestic market from competition in finished goods. There were occasional interruptions, as in 1957 and 1961, when export recessions necessitated some dampening of domestic demand, but the government had apparently learned how to cope with such problems. Farmers would produce exports; they would grumble about parasitical urban industry and services, but they generally accepted the

---

\* My reading of current policy-making derives from my study of New Zealand's economic history, see Hawke, G.R., *The Making of New Zealand: An Economic History*, (Cambridge: Cambridge University Press, 1985).

fairness of using part of their export receipts to build a more diversified economy and society. Some people took the rhetoric of egalitarianism seriously, but more were satisfied with the distribution of income which existed. The Arbitration Court relieved employers of most of their industrial relations problems and wage-earners were usually at least as content as farmers.

In the course of the 1960s this equilibrium was disturbed. Observers of international statistics and New Zealanders travelling abroad realized that the New Zealand economy was growing more slowly than many others. New Zealanders would not long remain content with a lower average standard of living than was available in European countries or in other countries which had been regarded as LDCs. The principal engine of economic growth in the international economy was widely seen to be international trade in manufactured goods. There was little chance of the European Community or other countries opening themselves to agricultural goods; even Britain was protecting its own farmers and attempting to join the European Community, and furthermore it was growing more slowly than many other countries. Therefore, the appropriate strategy for New Zealand was to join in the international trade of manufactured goods.

An early response was the New Zealand–Australia Free Trade Agreement (NAFTA), a limited step towards freer trade with Australia. Currency devaluation to parity with the Australian dollar in 1967 was an even stronger spur towards exporting manufactured goods to Australia. Another effort was the National Development Conference of 1969, which attempted to establish compatible objectives among different industries and to shift the economy towards a strategy based on exports while preserving consensus on income distribution.

This exercise was first knocked off course by the boom of the early 1970s. Buoyant export receipts seemed to remove the perennial shortage of foreign exchange. This was due to favourable international commodity prices and to the decision by many farmers that the favourable ratio of output prices to input prices which resulted from devaluation would not persist, so that it paid to concentrate on meat rather than on the longer-term returns from wool. There seemed to be no reason to worry about consistency and optimality of sectoral developments when imports could be readily financed.

A more serious blow to the strategy of the National Development Conference was the oil crisis of 1974. It became much more difficult for New Zealand to grow through exports when a large slice of world income was transferred from the markets for New Zealand exports to oil-producing countries.

New Zealand's response was to seek to spread the needed adjustment over several years by borrowing. This was in line with international recommendations to "recycle petrodollars". Although this strategy was probably unavoidable in 1974-75, since it is very hard to turn around an economy and a public sector from a situation where there seemed to be no cash constraint to one where real average incomes were markedly reduced, it was relied upon for too long a period.

The external environment was far from congenial. The government responded with the "growth projects", particularly after a further rise in the relative price of oil in 1979 made it sensible to use New Zealand's hydrocarbon resources which had accordingly become more valuable. It is accordingly understandable, but regrettable, that this action was presented in the readily-understood but misleading rhetoric of increasing New Zealand's self-sufficiency rather than as an element in a positive adaptation to a changed world economy. The real key was the competitiveness, in the light of international prices, of New Zealand's industries whether they were growth projects, suppliers to growth projects, or only indirectly linked with the growth projects.

At the same time, reduced real incomes intensified sectional conflict and fed inflation within New Zealand. Borrowing continued and was defended as a sensible way to develop New Zealand's resources, but in fact it was supporting domestic consumption at a level above that justified by production. Assistance to exporters was extended to the agricultural sector with the result that producers were given no incentive to look at trends in world markets. At the same time much of this support was captured by urban suppliers of inputs to farmers. Some farmers did gain, but less by increasing output of marketable produce than by selling farms at prices which included capitalization of government support.

There were, however, achievements during the 1970s. Exports of manufactured goods to Australia and other markets did grow. Some in the agricultural sector developed markets in horticultural products and kiwi fruit. The transport sector was significantly rationalized. A Closer Economic Relations (CER) Agreement with Australia greatly extended the NAFTA and provided exciting new opportunities for manufacturing. But by 1984, it was clear that existing policies were leading New Zealand into increased overseas debt while not removing its incomes disparity relative to other countries. Many people realized that radical change was needed. The Planning Council's Economic Monitoring Group observed in December 1983:

> The issues the Monitoring Group are addressing are sometimes summed up by saying there is a persisting gap between payments and earnings of foreign exchange. The gap is the result of the ways we organize and manage our economy. Simple solutions to close the gap, such as raising the price of foreign exchange (i.e., devaluation), are unlikely to be fully effective as long as we continue to organize and manage our economy the way we do. The traditional solutions of subsidising exports and protecting local manufacturing have been shown to be equally inadequate. Attempts to increase the earnings of foreign exchange by expanding production for export, lead through higher incomes to increased expenditure on imports, and hence to failure to close the gap. Similarly, attempts to expand production from import replacing industries also lead through higher incomes to increased expenditures on imports and hence—surprisingly as it may seem to many people—to failure to close the gap. Raising the price of foreign exchange will probably ultimately be necessary in a permanent solution to close the gap but this only makes sense if our economic organizations and management themselves work towards closing, rather than maintaining the gap.[1]

## 2. A New Policy Environment

The need for change was widely recognized. Most important, it was apparent to leading official economists. The election of a new government in July 1984, especially one which resulted from a snap election and had therefore given few hostages to particular interest groups, provided an opportunity for an unusually sharp change of direction.[2]

The centre-piece of economic policy became efficiency and equity. When official economists, who were allowed so little influence on immediate decisions in the early 1980s, had time for academic thinking, they reconsidered the traditional statement of the objectives of economic policy in terms of income growth, income distribution, minimization of inflation, balance of payments equilibrium, and full employment, and distinguished more clearly between those which were fundamental objectives and those which were means to those ends. Emphasis on efficiency and equity was the result.

Efficiency and equity immediately provoke in most people's minds the notion of a conflict, or in economists' minds, the idea of a trade-off. But they should equally provoke among economists the idea of a Pareto optimum, or, in ordinary language, the importance of ensuring that any choice is properly defined. That is, before a choice is considered, one must ensure that all moves which contribute to both efficiency and equity have been exhausted. Consequently for the first time for many years, policy formation began with a re-examination of what is equitable.

This is one of the powerful thrusts behind several policy initiatives since 1984. Does a system of import licensing create an equitable distribution of consumer goods in short supply, or even of materials used by industries? Does a system of constrained interest rates result in an equitable distribution of funds among competing would-be borrowers? Does a system of award rates supplemented by negotiated supplements create an equitable distribution of wages? There is a danger that one can create a series of apparently rhetorical questions which lead to the conclusion that everything erected over the preceding 40 years (or longer) should be swept away. Such a view is too simplistic, but these questions are worth asking. In the 1980s several institutions and practices that were created in the interests of equity are no longer serving that purpose.

There will eventually be some trade-offs between efficiency and equity. A modern civilized society cannot leave some of its members without an adequate standard of living; this necessarily creates some disincentives to providing for oneself. For example, if we rely on an insurance-based medical scheme, we still have to provide for those who gamble on good health and lose, and even for those who deliberately invoke moral hazard and choose to take the risk of relying on public support. Or if we seek to provide income support for those aged who are unable to support themselves, we affect the incentives of others to save. These tensions are unavoidable and guarantee that there must be continuing political debate about how the balance of advantage for society as a whole is moved by economic and social change. There is no prospect of redundancy for economists who seek to define and ameliorate poverty traps or for politicians able and willing to make the big decisions.

The new policy environment deals with both efficiency and equity. There is room for debate about each of the two components. It might seem that efficiency is a more "technical" and therefore a less debatable topic, but before one can determine whether one is securing maximum output from given resources, one must be sure of what is being produced. If broadcasting is simple provision of entertainment, then its efficient organization is very different than if it is "public broadcasting" (which is different from "commercial broadcasting") and is concerned with nation-building as well as entertainment. One must also know the market for which output is being produced. Whether or not a producer or marketing board is to be preferred to competitive marketing on efficiency grounds depends critically on the nature of the market being served. There are therefore debates about the meaning of efficiency.

Passions are even higher when the meaning of equity is debated. As the government put more emphasis on social policy after it was re-elected in 1987, divisions within its ranks became deeper, culminating in the December 1988 departure from Cabinet of the Minister of Finance who had directed the change in economic policy from 1984. Yet, throughout, efficiency and equity considerations were closely intertwined. In the case of labour market reform the Economic Monitoring Group argued:

> Both the efficiency with which resources are used, and the objectives of government intervention necessarily involve social as well as narrowly economic dimensions. A fundamental assumption on which our analysis rests is that it is possible to look at our labour market policy as concerned primarily with issues of economic

adjustment. That involves many social issues, such as the appropriate valuation of different kinds of work and the community's decision on minimum wage levels. But it does mean that matters of income distribution are primarily to be handled by taxes and benefits rather than through the wage system. This reverses a long-standing feature whereby Australia and New Zealand differ from most OECD countries in that our welfare systems have been little concerned with those in employment.

The argument for this is quite simply the social change which has occurred in New Zealand. It is no longer correct to think of most employees as men supporting families. It is not practicable to set wage rates with any particular family size in mind as was the policy of the Arbitration Court in earlier years. Wages have to be related to individuals and the value of the work they do. Income distribution remains a proper subject of debate, but it is better related to households than to individuals, and to the government's welfare policies than to employers. There are problems with this approach. Trade unions have always been part of a movement, a movement with social objectives wider than negotiating with employers over wages. In the last wage round, our trade unions demonstrated more capacity to influence the incomes of relatively low-paid workers than most commentators expected. To increase the separation of wage negotiations and income distribution issues, unions may have to be given a more influential voice in the formulation of policies about the latter. Unions are also concerned that with greater reliance on welfare schemes, low-income households may suffer from an unsympathetic future government. But that is an argument for participation in political processes rather than against separation of labour market and income distribution policies. There are also arguments that welfare benefits are less dignified than employment incomes. But payment of wages above the value of the labour which produces them merely disguises the element of subsidy, and recent social change has already vastly reduced any stigma associated with welfare schemes as the reception of national superannuation showed. The Economic Monitoring Group therefore concludes that it is reasonable to look at labour market policy primarily in terms of economic adjustment.[3]

The Economic Monitoring Group may have been somewhat ahead of public opinion in suggesting that transfer payments now attract little stigma, but the broad point can still be supported. There was a time when acceptance of national superannuation was an indication of failure; it is not now. There is still a belief that a family should be able to support its children, but it is reasonable to expect a gradual acceptance of the notion that society has to support young children. The notion that tax contributions justify social support in old age will be paralleled by the view that society has an interest in ensuring that young children do not experience relative poverty as society adapts to the two-income household. The Family Support package is more than an optional extra, added on to economic policy; it is an essential adjustment of policy so that economic and social policy initiatives can proceed hand in hand.

## 3. Processes

New Zealand has therefore begun a process of policy reform. Power is being shifted. Established pressure groups have lost influence. Patterns of public service influence on policy have been changed.

In the first term of the present government, from 1984 to 1987, the processes by which this was achieved seemed to cause only a little concern. The Minister of Finance directed the reform of economic policy with the support of his Cabinet colleagues. The thrust of removing subsidies from agriculture and industry, reducing tax rates while extending the tax base, and separating the commercial and regulatory functions of government, were supported by a united government. There were worries about the implications of reform of State-Owned Enterprises (SOEs) and the reform of traditional doctrines of ministerial responsibility.[4] There were also some concerns that a small group of public servants in the Treasury and the State Services Commission had gained undue prominence in official policy-making. And there was certainly conflict over the terms of reference and personnel of a Royal Commission on Social Policy, but these debates were mostly confined to the cognoscenti. Furthermore, it would generally be agreed that the process used to introduce a value-added tax called Goods & Services Tax (GST) was a model of public administration.

It drew on a wide range of advisers, including the early work directed by the Institute of Policy Studies; proceeded through the announcement of government's intentions in Green Papers and detailed consultation with interested parties and reports from specially constituted working parties of experts in particular fields; moved to White Papers which adopted many of the suggestions which resulted from that consultation; and finally resulted in legislation and implementation. A very complicated process which, while not without difficulties, went remarkably smoothly.[5]

There were significant changes in the influences on policy. In some fields ministerial decisions became less significant. The legacy of import licensing was that business leaders regarded government approval as essential for any major decision and several were surprised on informing the Minister of Trade and Industry of their intentions to be told that their initiative seemed admirable and that the government had no role other than to wish them well. Furthermore, especially in the tax reform field, the government turned to people knowledgeable in the fields of most relevance, rather than to established corporate organizations. The Manufacturers' Association, the Employers' Federation, the Federated Farmers and the Federation of Labour were somewhat by-passed.

The government's second term has been marked by more controversy. There is a sense in which the achievement of the first term can be said to be removal of subsidies and reform of the tax system. Reduction of industry assistance, both at the border and internally, promoted a better resource allocation and directed attention away from seeking privileges towards real economic activity. The tax system was changed so as to have lower rates and a broader base, which permitted a reduction of the budget deficit and better control of the level of public debt despite real growth in government spending, especially on salaries in the health and education areas. (There is remarkable difference between the common perception that the government has reduced taxes and cut expenditures, and the reality of reduced tax rates and a better balance between revenue and expenditure.) In the second term, there has been much stronger debate over how better returns are to be secured from public expenditure.

At a political level, this has resulted in debate over the future of social spending. The Minister of Finance attempted to take further steps towards lower tax rates on a broader base, and greater transfers to low-income groups, using a Guaranteed Minimum Family Income scheme to ensure that there were adequate incentives for finding employment. It involved a shift towards public funding rather than public provision of social services, and a withdrawal of public spending in some areas. A significant number of Labour Party personnel were not convinced that this strategy was right, and there was bitter debate at Cabinet level over how the process of reform should extend to social policy. The Cabinet division was resolved in December 1988 by the departure of the Minister of Finance. The 1988 Labour Party conference established a mechanism for cabinet-party consultation before the government departed from party policy; its adequacy remains uncertain.

At an official level, a wider range of personnel has been involved at the centre of government policy. One of the reasons for the Prime Minister's eventual withdrawal of support from the proposal of the Minister of Finance was that officials in departments other than Treasury and the State Services Commission questioned the extent to which that proposal was soundly based conceptually and statistically. In earlier years, it would have been expected that such disagreements should be resolved before papers went to Cabinet. During the first term, the forms of interdepartmental consultation which had evolved over many years were allowed to decay as ministers, suspicious of a conservative and bureaucratic public service, worked with particular officials equally committed to the reform of economic policy and sometimes instructed them not to consult with their colleagues. The decay proved to have costs.

The Cabinet Social Equity Committee became the focus of policy debate. It employs ad hoc Working Groups, composed mostly of representatives of relevant departments, but outsiders are also brought in.[6] Getting outsiders with appropriate expertise and available time is not easy, and there is now a trend for working parties to consist only of officials and therefore to resemble earlier officials' committees. A more significant departure from earlier interdepartmental officials' committees is that the convener of each group is charged with responsibility for a report. Departmental dissents must be recorded in the reports, but the working groups are not required to seek consensus. The intention is to have relevant issues sharply focused for cabinet ministers and to prevent the momentum of change being lost in interdepartmental disagreements. Reports have been presented on child care, post-compulsory education and training, occupational licensing, "comparable worth" and employment policy. On slightly different tracks, the Cabinet has also been faced with major issues in the fields of race relations and resource management. While this mechanism has clear advantages, some senior public servants see it as enabling ministers to bypass them for handpicked advisers, whether officials or others. And the Cabinet Social Equity Committee, even with reports from Working Groups, has not found it easy to reach decisions on what are necessarily difficult issues, both conceptually and politically.

Furthermore, the process of public consultation has not worked smoothly. In the field of superannuation or income support for the elderly, the government's proposals were widely regarded as unworkable and doctrinaire by those most involved and the government was less willing to use consultative groups and adopt their suggestions. What was initially announced as firm policy in a White Paper was later presented as proposals in a Green Paper and a great deal of uncertainty was generated. More generally, interest groups have not readily accepted that cabinet ministers need processed material for their decision-making, and have demanded consultation on the reports of the Working Groups, not only on matters of implementation but through re-litigating earlier debates. And, as some business groups realized that the government's objectives differed from theirs, the generally favourable media response to the government dissipated. The government is not committed to a diminution in a role of the state; in its first term, the implications of "improved quality" in public expenditure, such as government sought, coincided with support for restraining government expenditure as business groups sought, but in the second term, the government has resisted withdrawal from public funding and activity.

It could be argued that most progress has been made in education and health administration and that the major steps were taken before the Cabinet Social Equity Committee was given its present prominent role. In the case of the school system, the Picot

Committee was an ad hoc body which secured wide public acceptance, and the government has proceeded through a White Paper, *Tomorrow's Schools*, and seems to be holding firmly to its course despite counter-attacks. In the case of health, the responsible minister proceeded on his own agenda (which included a great deal of continuity with earlier changes), ignoring much of the analogous Gibbs Report. But it can equally be argued that the main problem is that the issues are complicated and that achieving an appropriate reform of social policy is something that can occur only with vigorous debate within the government and among the public. Reform of early childhood and tertiary education, directed by the Cabinet Social Equity Committee, seems to be following the same path as reform of the school system.[7]

## 4. Tax Policy[8]

One broad area in which the government moved with great speed was tax policy. Serious commentators had long known that the common view that taxation in New Zealand was especially heavy was erroneous; what was unusual about New Zealand was that taxation had come to be concentrated in the form of personal income tax. As long ago as 1967, the Ross Committee recommended a shift towards indirect taxation, but the political difficulties of such a move were then regarded as insurmountable. In the early 1980s, the McCaw Committee repeated the recommendation, but while income tax reductions were made, there was no serious move on direct taxation. A budget problem was the result of this inaction.

From 1984, the new government moved promptly towards a wider ranging goods and services tax (GST) and accompanying changes in other taxes. The administrative difficulties of GST were immense, but so was the determination to deal with them, and it can now be said that the introduction was extraordinarily smooth. The political difficulties were overcome precisely by a coordination of economic and social policy as discussed above. On its own, the introduction of the GST would have had undesired implications for income distribution; but GST was not introduced on its own. The accompanying moves to family care and then family support provided compensating changes in income distribution.

Tax policy reform involves much more than the GST. The basic thrust has been to extend the tax base and reduce rates of tax and opportunities for avoidance. The process is a continuing one. The intellectual problems of dealing fairly and effectively with income derived from overseas are far from solved; neither are those concerned with income which can be taken in the form of capital gains. Equally pressing is the appropriate treatment of savings, especially those in the form of superannuation contributions. Debate about whether the tax base should be extended to include wealth is in a comparatively early stage.

What we can say with confidence is that issues of tax policy have been approached with a genuine attempt to promote efficiency and equity. Furthermore, they have been approached at a high level of intellectual commitment. Policy-makers have attempted to look beyond the superficial first effects of any particular measure and to include in their analysis the ultimate effects on the way in which resources in New Zealand are used (thus confronting the fundamental issue of incomes in New Zealand relative to those available elsewhere) and on the distribution of income, not only through the initial impact, but after allowing for reactions to government initiatives.

Professional debate tends to be directed towards the efficiency effects of various taxes. But the wider debate has an equally strong concern with equity. It focuses on the issue of progressive taxation. The number of steps in the income tax scale has been progressively reduced since the late 1970s. The McCaw Committee of the early 1980s was especially influential in endorsing the direction of change. In particular, by looking at the incidence of tax on both individuals and households, it made a cogent case that exempting low incomes from tax was not necessarily progressive in its effects. All tax payers benefited from the low rates of taxation on the initial income tranches, and some low individual incomes contribute to households with relatively high overall incomes. There are good reasons for believing that New Zealand's taxation system has not been as progressive as rhetoric would suggest, and that income maintenance schemes would serve their purposes better if there were more emphasis on directed expenditure and less on differential taxation. Nevertheless the idea of progressive taxation as an element of social justice is deeply embedded. It has been present in New

Zealand taxation since at least 1891. Many people find it hard to accept that tax incidence does not depend only on a progressive tax scale. At a superficial level, there remains a considerable body of opinion that any social expenditure can be financed simply by increasing taxation of the rich, but even when that is put to one side, intellectual and political divisions ensure that tax policy remains a topic of serious debate.

## 5. Regulatory and Institutional Reform[9]

The central thrust of government policy is the effort to establish a stable medium-term framework within which producers and consumers can choose to arrange their own affairs. This provides a field-day for some journalists and critics who think that it entirely precludes "intervention" or "regulation" and see any economic action of the government as inconsistent with its declared policy. But the real policy is to decentralize decision-making so that decisions are made where information is most readily available, where people have most commitment to making their decisions effective, and to people who will bear most costs if their decisions are inappropriate.

The government's policy has been "regulatory reform" rather than "deregulation" but, as it started from a position where official controls were widely regarded as excessively detailed, it has not been easy to maintain that distinction in popular discussion. Regulatory reform in the financial sector consisted of sweeping away a great many controls that had little economic support, and which were administratively impracticable in the face of changing technology.

The government has revised the Commerce Act and, although there are many complaints from the business world about how it is operating, those complaining generally have specific decisions in mind—those which happened to go against them. Nevertheless, it continues to be difficult to find a system of rules for dealing in securities which is compatible with both economic logic and legal ideas and procedures. The government has also established an Economic Development Commission and it has been looking at the general area of regulation reform, trying to find an acceptable balance between the need for government to act quickly on occasion and the desirability of deliberate and open scrutiny of proposed regulations.

An important aspect of the Economic Development Commission is the attempt to create "transparency". The argument is that if there are reasons for government facilitating the operations of some group in society, they should be exposed for public scrutiny and debate. Policy should owe more to argument and less to the privileged access to decision-makers, political or official, by particular interests.

The same notion has informed moves on industry assistance generally. The government has generally won high praise for its "micro-economic" policy, by which it is meant its actions on protection and assistance to individual industries, and its abandonment of regulatory restraints, especially those on the financial sector. Naturally, even here, the verdict is far from unanimous. Farming interests believe that the removal of assistance has not been even-handed, and those formerly employed in industries adversely affected by the removal of protection tend to believe that the government has made a wrong judgement about the social value of their activities.

It is natural for an economic historian to reflect that this is not the first time that deliberate withdrawal of the government has been part of a radical reform program. When Adam Smith advanced the analysis of *The Wealth of Nations* he was not, as is often thought today, acting as advocate for business interests. He was arguing for a diminution of privilege, for the removal of access to public funds by the aristocracy and their protégés (of which, ironically, he was one himself). He wanted to prevent unwarranted public interference in the interests of those whose efforts would contribute to public wealth. His view of business interests was that they would indeed "conspire against the public" so that institutions should be arranged to minimize their opportunities to do so. That was a radical lesson, and it is one that should be relearned from time to time.

Furthermore, reliance on "market forces" had a positive social aspect. It was that individuals should be free of the constraints of outdated social conventions and able to pursue their own interests within a framework which prevented, as far as possible, one individual from damaging another. Relying on the market meant removing the constraints of aristocracy and unwarranted social deference. Now that the market is often portrayed as inimical to qualities such as cooperation and humane feelings, it is worth occasionally remembering the origins of the idea, and the desirability of permitting individual initiative in an anonymous fashion.

The policies which have been adopted in New Zealand owe a great deal to analyses of New Zealand's own position, but they clearly relate to an international climate of ideas and policy changes in other countries. They are sometimes described as "Reaganite" but more often as "Thatcherite". There are both differences and similarities between the policies which have been followed in New Zealand and those pursued in Britain. The differences revolve around motivation. The New Zealand government has not sought a diminution of the state's role for its own sake, although some people have argued that it should do so. Nor has the New Zealand government espoused any general return to the supposed family virtues of an earlier age. The similarities are that efficiency is promoted by a clear separation of different kinds of objectives. When a government department was invited to pursue both social and commercial objectives, it could arrange things to suit its own convenience, claiming that commercial objectives had to be tempered by social considerations or that social objectives had to be sacrificed for commercial pressures. The government tried to separate state commercial activities and organize them as state-owned enterprises with clear commercial objectives. The essence of the SOE reorganization was to facilitate monitoring of efficient use of assets which are owned by the public, entrusting commercial objectives to management, while reserving social considerations for the government which can contract with an SOE where social and commercial considerations conflict. This has the additional advantage of making "transparent" the use of public funds for particular purposes; it will be easier to distinguish considered responses to social needs from compromises with special interests.

Debate about the SOE reorganization has moved on to questions of whether trading assets should be privatized as the government wishes to reduce its debt level (and free itself from interest payments) by the sale of public assets.

The most importance issue related to debt is whether the price offered for public assets is greater than the discounted value of the income stream which the government can expect to earn from those assets. The corporatization process was directed towards improving the return earned from public assets, and the first issue now is whether the private sector thinks that it can earn more than state-owned enterprises. Some people think that private enterprise is always more efficient than public sector management. They argue that the implicit government guarantee on state enterprises, the belief that state enterprises will not be allowed to fail, reduces the incentives and imperatives on management and ensure that efficiency will eventually decline. They also argue that there is a greater likelihood of political interference with SOEs than with private companies; that SOEs will not be allowed to manage their businesses in a commercial way but will be required to make political compromises. The international literature is more mixed than such people sometimes claim, and includes cases of efficiency advantages for both public and private enterprise, with a majority probably leaning in favour of private enterprise but with the clearest conclusion being that generalization is difficult. It would, however, be right to say that the literature includes no significant support for a general preference for public enterprise.

It is immediately clear that a policy of selling state assets requires the government to have a capacity to assess the value of the future income stream it can expect from those assets in order to compare it with the price being offered by private purchasers. The exercise is demanding in terms of economic and financial expertise.

The issue of the appropriate selling price is not the only one. The government must also take an interest in the future plans of purchasers of public assets and seek to decline offers from asset-strippers even though the future intentions of purchasers are obviously difficult to ascertain. There is also concern about national sovereignty. The government's position is explained as having three major components. First, national sovereignty is diminished when debt is incurred, not when assets are sold. The government's freedom to manoeuvre is restricted by debt obligations, and the sale of assets should be seen as a reorganization of government's financial position and of the influence of people overseas, not as a simple extension of the latter. Second, the government wishes to secure the best possible bargain for the community as a whole, and that requires it to secure the biggest gap possible between the asset price and the discounted value of the income stream available from public management. Selling to New Zealanders alone would diminish that gap, and there could be no guarantee that further sales would not result in overseas ownership anyway. Restricting sales to New Zealanders would therefore merely provide a gain for selected New Zealanders rather than to tax payers in general. The government could secure electoral advantage from distributing public assets to selected investors in New Zealand, but that would be contrary to its intention of guarding the public interest. Thirdly, national sovereignty rests much more with the government's power to legislate and regulate than with ownership of assets.

There are various contrary arguments. There is a general argument about whether power goes with ownership and whether "New Zealand ownership" in a world of international interdependence has any value for political and social decision-making. Another depends on identification of characteristics of particular public assets which has given them "strategic" importance such that direct government control should be retained. This has to be pursued on a case-by-case basis. Both go beyond the question of overseas ownership. It can be argued, for example, that what is important about the ownership of the Bank of New Zealand is the possibility of conflict of interest of directors and owners and not whether they are New Zealanders or not.

Sales of public assets pose problems for private investors in New Zealand. The local capital market has been accustomed to a small range of instruments for which it is experienced in assessing relative balances of risks and rewards. It faces a large learning process as loan and equity finance is sought for a new range of public and private entities. There is a great deal more to the concept of the ability of the New Zealand capital market to "absorb" a program of sales of public assets than is generally realized.

All of these issues go well beyond the common discussion in terms of the government's level of overseas debt. The economics of privatization are essentially matters of efficiency. There may, however, be parallel forces which are common in popular discussion concerning the government and financial reconstruction of companies. If overseas creditors and rating agencies insist on using similar criteria for government debt as for corporate debt, then an exchange between private financing and government financing may be significant. A process which includes all government asset sales, repayment of government overseas debt, and the private overseas borrowing to finance the purchase of the assets may well be described by economists in terms such as "fundamentally irrelevant" unless it results in an improved use of the assets. Accountants are more likely to regard debt-equity ratios in the government's account as significant, and rating agencies and overseas investors may also be more likely to do so in accounting terms. There would then be an effect on credit ratings and on the risk premiums included in interest rates.

Discussion of the appropriate regulatory regime for privatized enterprises shows how far scepticism of regulation has proceeded in recent years. It is interesting to reflect on why this has happened. In some cases, technology has changed and former means of regulation have been rendered inappropriate. There was, therefore, necessarily some new thinking about what regulation is required and how it can best be administered. But there has also been an increased emphasis on the efficient use of resources. This has directed attention to whether existing regulatory agencies are efficient in their own use of resources, and whether their instruments are well designed for securing declared objectives without undesirable side-effects. (The economic issues are, of course, intertwined with political arguments, both about the desirability of individual freedom in general, and about the desirability of social objectives other than maximizing income.)

A good deal of attention is therefore now directed towards defining exactly where a monopoly exists and means of ensuring that its power is compatible with social objectives. The principal concerns with this strategy of "ring-fencing" well defined monopolies are that fences may not be permanent and that we need to define equally carefully the mechanisms which permit appropriate monitoring of the use actually made of monopoly power.

Issues of the organization of state trading activities have therefore become assimilated with wider questions about the appropriate role of the state. That has directed attention towards the appropriate organization of the core public service.

The present range of activities in which government is involved and for which it employs public servants can be thought of as a continuum from things which are necessarily the prerogative of government to those which are potentially contestable by the public and private sectors, to those which are clearly contestable, and on to fully commercial activities and eventually to those matters which are held, virtually unanimously, to be best left to households and private organizations. Of present government activities, policy formulation and regulation belong at one end, service delivery is mostly in the contestable area, and the SOEs are wholly in the contestable area. The precise placing on the continuum of health and education services, for example, is something on which policy debate is now focused.

Various aspects of the organization and management of the core public service are being considered within such a framework. For example, in considering how accountability should be achieved, ministerial involvement in planning, and control agency, qualitative evaluation is much more likely to be required in those areas which are considered to be the prerogative of government; while as one moves towards more commercial concerns, some kind of board management and quantitative market-related performance criteria are likely to be appropriate. In pay-fixing, occupation-based bargaining structures are likely to be more important in the least contestable areas while enterprise bargaining is appropriate in commercial activity.

The principal driving forces in the reform process are perceived needs to clarify objectives and delivery systems, improve accountability mechanisms, create greater flexibility and adaptability in personnel practices, and reduce the role of control agencies in the operations and departments. Current thinking is clearly intended to preserve a unified public service in core functions while providing more freedom from uniform practices in different policy and administrative functions.

There is inevitably a great deal of concern about their future expressed by existing public servants. There is a clear thrust in the State Services Commission towards finding appropriate places in the government's organizations for all of its staff, enabling managers to transfer staff to where they can best contribute to the government's objectives without abandoning the principle that dismissal is a last resort. Nevertheless, many people have lost their jobs, and expensive redundancy packages do not always seem adequate to them.

There are also wider questions which need further reflection and innovative thinking. The relationship of the accountability of public servants to the accountability of the executive to parliament is one of the most important ones. Questions about the relationships among departments remain problematical. Attempts to improve public sector financial management and accounting involve more monitoring and raise questions about the role of Treasury. The creation of a Senior Executive Service within the public service, a group of skilled and experienced policy advisers and managers able to move from one department to another, does not sit entirely comfortably with the emphasis on departmental autonomy.

Nevertheless, it is hard to see the process of change now under way being stopped. There is at least implicit in what has been done, a framework which provides for a wide variety of specific arrangements within a coherent whole. A State Sector Act has established new relations between ministers and chief executives of government departments, and a number of departments have been reorganized.

## 6. Labour Market

In the case of general economic policy, the Economic Monitoring Group argued that a significant and sharp switch was needed to demonstrate that the rules of the game had changed and that people would be better occupied responding to a new situation than attempting to preserve existing assistance.[10] In the case of the labour market, the important matter was to maximize the chances of getting change in the right direction, moving towards the right amount.

There are strong links between the key elements of the framework of the labour market: the registration system which essentially reserves coverage of employees in particular occupations to individual unions; "blanket coverage" which extends the wage rates of awards to employers not directly involved in their negotiation; compulsory conciliation; and arbitration. Although these links make it difficult to effect change in any one while leaving the others unchanged, the Economic Monitoring Group favoured an evolutionary approach to reform.[11] In its Labour Relations Act of 1987, the government adopted the same view. Compulsory arbitration was abandoned, but the strategy adopted was to widen the possibilities for negotiation between employers and employees, leaving them both to discover how a less constrained environment could work to their mutual interest. The Act has been much criticized by both employers and trade unions. But the Act has the potential to stimulate desirable change, the key being a change to more emphasis on bargaining.

Current trends in international thinking put a great deal of emphasis on management within the firm as the source of productivity advance, and that in turn is the source of increased incomes which have been identified as one of New Zealand's key needs. "Management" here is to be understood very widely, to include communication and cooperation at all levels of employment. The essential test of the Industrial Relations Act is that if employers are right that they are able to offer better working conditions as a result of negotiating with their own employees, the mere availability of this possibility should make a significant difference in the way in which awards are negotiated. Where present unions are ineffective, they would be replaced by new or re-organized unions. There are signs that this is happening, although there are obviously still some doubts among unions over whether matters like regional differences are being accommodated fairly and by some employers over whether the pace of change is fast enough.

This is given urgency by levels of unemployment which are far above what New Zealand has experienced in recent decades, although not markedly out of line with those of other countries which have experienced a disinflation similar to that experienced in New Zealand in the 1980s. The Government's reaction has been to put less funding into subsidized employment schemes, believing these to shift rather than reduce unemployment, and to direct funding towards training schemes which have some prospect of improving employability. The case for such a shift from "passive" to "active" policies is strong, but it is unlikely, even in the medium term, to produce an employment balance which would be widely judged as acceptable. Most economists believe that the only available approach to reducing unemployment more quickly than can be expected from a continuation of current trends and policies is through further deregulation of the labour market. Trade unions believe, however, that such arguments are merely a disguise for a wish to destroy the trade union movement in the interests of business profits. They argue for direct government stimulation of investment. The issue of unemployment then becomes part of a wider debate about social policy.

## 7. Social Policy

In its first term from 1984-87, the present government concentrated on getting in place a change of economic policy. Because of the joint objectives of efficiency and equity, that inevitably provoked debate over social policy. In describing tax policy, state sector reorganization and labour market policies, I have tried to give an impression of the way in which pressures which are often described as "social" were central to defining the economic problems facing New Zealand and in the government's response to them.

It was, however, always clear that a more concentrated reformulation of particular aspects of social policy was on the agenda after immediate economic issues. After a great deal of hesitation, the personnel of the Royal Commission on Social Policy was agreed on and it reported in April, 1988.[12] There was a consensus on the general desirability of a wider ranging review of the social policy which had developed from the Social Security Act of 1938, but no agreement on how it should proceed. The Commission adopted the process of gathering a very wide range of opinion and it compiled a long document which will be a source for academic study for many years. But it did not develop a sharp analysis of where social policy should now be directed. Popular opinion is naturally conservative—people tend to be satisfied except for their particular concerns—and to be divided. Thus people could readily agree that social and economic policy should be integrated, but they did not recognize how integrated they are already, and where there are sharp differences of opinion, the government was given little guidance on which option it should take.

Decisions had to be taken while the lengthy process of a Royal Commission continued; dissatisfaction with an appearance of preemption contributed to the Royal Commission shortening its timetable and making an early final report in 1988.

The government has therefore reverted to task forces to define the social policy issues it needs to address in a similar way to that in which it proceeded on tax policy, and as discussed above, sought to coordinate them through the Cabinet Social Equity Committee. The thrust of policy is towards securing "quality" of government expenditure, ensuring that public assets and public funds are used as well as possible for the objectives towards which they are directed and that those objectives still reflect the overriding objectives of efficiency and equity. It is, however, difficult to persuade people that that is different from reducing government expenditure for its own sake. People tend to think that fiscal pressures can easily be resolved by reducing spending other than that which is directed towards their particular interests.

The intellectual and political difficulties are great. Economists think naturally of income maximization, not because they are materialists but because incomes provide people with choices. Even for policy objectives which are usually expressed in terms like "participation in society" and which are therefore thought of as "social" rather than "economic", the experience of most people depends on their ability to make decisions about the disposition of household incomes. This is especially true of incomes earned in the labour market, which is a principal reason why employment levels have such social significance. But participation in society extends to collective activity, to the sense of belonging or social cohesion which follows from sharing in joint endeavours. This may relate to something as significant as access to "free" medical care, or to much less obvious arrangements such as ability to make telephone calls without identifiable costs. Society may want to allocate some goods or services according to "need" rather than according to income. It then has to find some means of judging "need" and some process for ensuring that responsible institutions satisfy their mandate rather than use resources for their own purposes. Economists would be wrong to reject any such possibility without further discussion. Departures from markets with cash transactions by individuals are not necessarily signs of inadequate enlightenment calling for further doses of economic doctrine.

Nevertheless, it is not enough to say "collective action" and rule out further debate about appropriate policies. Costs have to be allocated whether it be by subscription to a club or by payment for individual services, and most of us have experienced debates in voluntary clubs about whether members are sharing fairly in costs and benefits. It is not always the individualist economists who undermine collective arrangements; those free telephone calls are being pressured by technical changes that mean the principal beneficiaries are business users of fax machines rather than individuals maintaining social contacts. Cosy collective arrangements do not always benefit the relatively powerless in society.

Nor does inflation, or borrowing so that the cost of current consumption is passed on to future generations. Monetary policy has been used to gain control of inflation and for the first time for many years, it is now less than that experienced by New Zealand's trading partners. The cost has been high because high interest rates have discouraged investment and employment. They have also attracted an inflow of foreign funds and its effect of raising the New Zealand exchange rate has strengthened pressure on producers of tradeable goods. But this could have been avoided only by a faster diminution of the budget deficit; the essential policy decision was one with as many social as economic dimensions: the relative share of adjustment costs to be borne by those dependent for employment and incomes on the tradeable sector and those similarly dependent on government expenditure.

Most social of all is the judgement that present New Zealanders should rely less on forcing succeeding generations to finance current consumption. Discussion of the budget deficit and the general strategy of fiscal policy is often regarded as technically demanding and of interest only to economists, but it is at the core of a social strategy.[13]

Indeed, while the government has been criticized for adopting an "inflation first" strategy, its overall policy has not had any single objective. "Efficiency and equity" has been taken seriously. Macro-economic policy has been guided by a desire to improve the government's use of the resources under its direct control and to guide private decisions towards a sensible use of a resources relative to the international economy. Within that, monetary policy has been directed to controlling inflation and to strengthening control of monetary trends and the exchange rate has been allowed to float. (It is difficult for people to believe that when the monetary authorities refrain from fixing the exchange rate, social control may be increased, but that is in fact the case.) At the same time, micro-economic policy has been aimed at efficiency.

The whole structure of policy is aimed at satisfying community aspirations, social and economic. But those community aspirations are diverse and inconsistent so that no government can ever be entirely successful. Criticism of the government has grown as unemployment levels have climbed and government policies have not been successful in making New Zealand an attractive place for investment with consequent employment expansion—inflationary expectations have consistently lagged behind what has been achieved, and uncertainty about tax and other policies has hindered investment. But the issue has not been the relative importance to be attached to inflation and unemployment, but whether the mix of policies can be adjusted so that employment is promoted. Any attempt to promote employment simply by increased expenditure would destroy the whole strategy and could be expected to provide no more than a temporary palliative while undoing what has been achieved towards resolving the problems which were so apparent in 1984.

## 8. Conclusion

There are many aspects of economic policy-making which I have barely touched on here. New Zealand continues to put a great deal of emphasis on international trade negotiations, especially on the treatment of agricultural products in the current GATT round. It continues earlier efforts to widen its trading area, in the development of the CER agreement with Australia and its possible extension to other countries such as Canada or Asian countries.[14] I have chosen to pay less regard to those areas, important though they are, because they have seen less change in recent years than domestic economic policy and its extension to social policy.

New Zealand has made a determined effort to face problems, an effort which has been controlled by the Cabinet but which has been open to consultation with a wider range of people than had been customary. Nevertheless, both the policies which have been adopted, and the processes by which they have been formulated remain subject to improvement, especially as social issues become more prominent.

## Notes

1. *Foreign Exchange Constraints, Export Growth and Overseas Debt: Report No. 1 of the Economic Monitoring Group*, (Wellington: N.Z. Planning Council, December 1983), p.2.

2. For alternative accounts of the Labour Government elected in 1984, see J. Boston and M. Holland, eds., *The Fourth Labour Government*, (Auckland: Oxford University Press, 1987) and R. Rabel, "New Zealand" in R. Baker and C. Morrison, eds., *Australia, New Zealand and the United States: National Evolution and Alliance Relations*, (forthcoming: a summary of the first workshop in a series of three is available as *Australia, New Zealand and the United States: Changing Societies, Politics and National Self-Images* {Hawaii: Resource Systems Institute, East-West Centre, 1988}).

3. *Labour Market Flexibility: Report No. 7 of the Economic Monitoring Group*, (Wellington: N.Z. Planning Council, June 1986).

4. The Institute of Policy Studies was influential in generating studies and discussion of such issues. John Roberts, *Politicians, Public Servants & Public Enterprise*, (Wellington: Victoria University Press for Institute of Policy Studies, 1987); Peter McKinlay, *Corporatization: the Solution for State Owned Enterprise?* (Wellington: Victoria University Press for Institute of Policy Studies, 1987); G.W. Jones, *Privatization: Reflections on the British Experience*, (Wellington: Victoria University Press for Institute of Policy Studies, 1987); Greg Parston, *The Evolution of General Management in the National Health Service*, (Wellington: Victoria University Press for Institute of Policy Studies, 1988); *The Producer Board Seminar Papers*, (Wellington: Victoria University Press for Institute of Policy Studies, 1988); John Martin, *A Profession of Statecraft*, (Wellington: Victoria University Press for Institute of Policy Studies, 1988); and *State-Owned Enterprises: Privatization and Regulation–Issues and Options*, (Wellington: Victoria University Press for Institute of Policy Studies, 1988).

5. Alex Texeira, Claudia Scott and Martin Devlin, *Inside GST: The Development of the Goods and Services Tax*, (Wellington: Victoria University Press for Institute of Policy Studies, 1986).

6. The author chaired one of these working groups which produced the *Report on Post-compulsory Education and Training*, (Wellington: Government Printer, 1988), which may contribute to the more favourable view expressed here than is held by several senior public servants.

7. The Picot Report was entitled *Administering for Excellence*, the Gibbs Report *Unshackling the Hospitals*, the White Paper on the tertiary sector following the Working Group Report cited above, *Learning for Life*, and that on early childhood education following the Mead Report to Cabinet Social Equity Committee, *Before Five*. All were published by the Government Printer in 1988 or 1989.

8. The debate about tax policy can be followed in Texeira, Scott and Devlin, *Inside GST;* John Prebble, *The Taxation of Controlled Foreign Corporations*, (Wellington: Victoria University Press for Institute of Policy Studies, 1987); Richard J. Vann, *Trans-Tasman Taxation of Equity Investment*, (Wellington: Victoria University Press for Institute of Policy Studies, 1988); Cedric Sandford, *Taxing Wealth in New Zealand* (Wellington: Victoria University Press for Institute of Policy Studies, 1987); and Toni Ashton and Susan St. John, *Superannuation in New Zealand: Averting the Crisis*, (Wellington: Victoria University Press for Institute of Policy Studies, 1988), and the references given therein.

9. An early stimulant of policy discussion was *The Regulated Economy: Report No. 5 of the Economic Monitoring Group*, (Wellington: N.Z. Planning Council, September, 1985), but much of the more recent public discussion has been led by the Economic Development Commission. See, for example, David Haarmeyer, *Competition Policy and Government Regulatory Intervention*, (Wellington: Economic Development Commission, October, 1988); *A Generic Approach to the Reform of Occupation Regulation*, (Wellington: Economic Development Commission, December 1988). The Law Commission has also been a prolific contributor.

10. *Strategy for Growth: Report No. 3 of the Economic Monitoring Group*, (Wellington: N.Z. Planning Council, 1984).

11. *Labour Market Flexibility: Report No. 7 of the Economic Monitoring Group*, (Wellington: N.Z. Planning Council, 1986).

12. *The April Report*, (Wellington: Royal Commission of Social Policy, 5 volumes, 1988). The Royal Commission also published many supporting papers, most notably *Working Papers on Income Maintenance and Taxation*, (March 1988) and *The Role of the State: Five Perspectives*, (March 1988).

13. *Tracking Down the Deficit: Report No. 8 of the Economic Monitoring Group*, (Wellington: New Zealand Planning Council, 1987).

14. Sir Frank Holmes et al., *Partners in the Pacific* (Wellington: N.Z. Trade Development Board, 1988).

# Philippines

## A Political Economy of Philippine Policy-Making

*Emmanuel S. de Dios**

### Introduction

Revolutions and their aftermath normally provide abundant grist for the research mill owing to the sweeping changes they entail and the stark (almost facile) contrasts with the old regime they invite one to draw. In this sense, the 1986 Philippine revolution must prove somewhat disappointing, and the study of Philippine economic policies, especially those pertaining to debt and trade, are fairly untypical. For here, the concern would seem to lie mainly in explaining the striking *continuity* in policies in these important areas, a continuity all the more remarkable since it persists against a recent backdrop of large-scale political upheavals and realignments.

This paper attempts to explain this post-revolutionary continuity in crucial economic policies by tracing the social and historical processes influencing both the agenda and constituency that carried the Aquino administration to power in February 1986. The hope is that by doing so, a framework may be developed which

---

\* M.F. Montes and R.D. Ferrer provided helpful discussion on the content of this paper, but the author takes sole responsibility for the views expressed.

will allow the unified treatment of the government's responses to both political and economic imperatives.

In Section One, a brief analysis of the political economy prior to and under the Marcos dictatorship is presented. In Section Two the aim is to locate historically the composition of the Aquino constituency, the content of its agenda, and the extent to which this has been changing. It is pointed out that shifting political imperatives have affected the government's decisiveness in implementing economic policies. Section Three discusses economic issues concentrating on the debate on debt strategies and interpreting the social significance of various viewpoints and lobbying interests arrayed on this question. A concluding section poses unresolved problems regarding the analysis of Philippine political economy.

## 1. Policy-Making Prior to and Under the Marcos Regime

In many respects, the installation of the Aquino government meant a reversion to political practices and rules prevailing before the imposition of martial law in 1972 (Nemenzo, 1988). If only for this, it is important to gain an idea of the status quo before the dictatorship. A further reason, however, is that the splits within the ruling elite engendered by the old political system were themselves an immediate cause for the imposition of the dictatorship, and resolving this schism is a problem that continues to confront the present government.

### *Premartial Law Politics*

The political system prevailing from independence in 1946 to 1972 might best be described as "elite democracy". As Lande (1964:1-2) observed some time ago:

> ... (T)he Philippine polity, unlike those of most present-day Western democracies, is structured less by organised interest groups or by individuals, who in politics think of themselves as members of categories, i.e., of distinctive social classes or occupations, than by a network of mutual aid relationships between pairs of

individuals ("dyadic" ties, in anthropological terminology). To a large extent the dyadic ties with significance for Philippine politics are vertical ones, i.e., bonds between prosperous patrons and their poor and dependent clients. The Philippine political elite is drawn largely from among those who can afford to be patrons, i.e., from landowners who have tenants, from employers, and from professional men whose occupations permit them to do favors for large numbers of ordinary voters. Members of this elite, ranging themselves under the banners of two national parties, compete with each other for elective offices.

An important omission in this observation is the closing of the circle: from wealth to political influence, to the preservation or expansion of wealth. It is not simply that wealth was required to gain entry into the political elite. More pertinently, even prior to 1972, access to the political machinery was in fact a principal means of direct or indirect wealth-accumulation, or as Ferrer (1988:70) writes, "the prime mechanism for the reproduction of property is a political mechanism". In Philippine political culture, this has been known as the perennial *problem* of graft and corruption.[1] It is probably indicative that the two major political parties prior to martial law alternately came into power and lost it on variants of only two basic issues, namely official corruption and political violence. The view that political (or generally non-economic) levers are key means for preserving and accumulating wealth allows one to explain some stylized facts regarding Philippine politics which would otherwise be traced to exclusively *cultural* factors. On this view, therefore, politics, in an underdeveloped context, is itself a major form of organization of the economy. Political violence may be viewed simply as one form of *investment*, and corruption as a form of *return*. Put in this perspective, the scale of corruption and violence becomes more comprehensible, though perhaps not more tolerable.

From 1946 until the late 1960s, elite factions were able to keep up a modicum of coexistence which took the form of electoral alignments and defections, dictated only by calculation and convenience, even as the major political parties alternated in power. The only significant change in this pattern was the re-election of Marcos as president in 1969.

## Elite Crisis

By 1970, however, three sets of circumstances combined to make matters come to a head. The first was a worsening economic crisis in the form of severe balance of payments difficulties provoked by heavy electoral spending during the presidential elections. This put a heavy strain on the currency, which was drastically devalued in 1970 under an IMF plan. The resulting inflation ate into real incomes, especially in the urban areas, and led to worker-student unrest. A second set of factors pertained to the uncertain status of U.S. assets in the country, owing to the impending termination in 1974 of parity rights given to U.S. citizens under the Laurel-Langley Agreement; this occurred against the backdrop of several decidedly nationalist Supreme Court decisions and bills in the legislature aimed at curtailing the scope of U.S. holdings. With the re-establishment of the Communist Party in 1968 and the formation of its armed component, the New People's Army, which drew its members mostly from the peasantry, in 1969, the nationalist protest movement acquired both a militant edge and a mass character. Finally, however, there was the outstanding political question of Marcos's term, which was to have ended in 1973 (a third term being disallowed under the prevailing constitution) but which he was maneouvring to extend under a constitutional amendment. The split in the Filipino elite had reached crisis proportions, owing mainly to this bid for political and financial hegemony by the Marcoses and Romualdezes (who were relative upstarts) as against the more established political clans such as the Lopezes, Osmenas, Aquino-Cojuangcos, and Jacintos. It was these whom Marcos, adopting a populist rhetoric, referred to as "oligarchs", to legitimize the expropriation of their holdings and their political persecution. In turn, it would be these same families, mostly coming from exile in the U.S., who would be restored to social and economic prominence with the coming to power of Corazon Aquino.

Domestically, therefore, a state of paralysis and deep division among the politically prominent landowning and business classes provided the condition under which martial rule and dictatorship would at least be tolerated—if not exactly welcomed—by other elite interests who were not as directly or as deeply involved in the conflict between the political heavyweights. Many of these clans (some of which, like the Zobel-Ayalas and Sorianos, were less politically active but financially prominent) would be coopted by the dictatorship and later have to contend with the stigma of collaboration.[2] Hence, while the dictatorship manifested a degree of independence of the traditional elite, it could not afford to alienate the entire stratum of landowners and business, and chose its targets carefully. In the initial years of martial rule, the economic targets among the elite coincided with the politically important ones and consisted of clans such as the Lopezes, the Osmenas, Elizaldes, and the Aquino-Cojuangcos. In this manner, the establishment of a dictatorship resolved, as it were by force majeure, the outstanding political issues among the Filipino elite. Similarly, for foreign and especially U.S. investors, as well as for the U.S. government, the advent of one-man rule meant an end to uncertainty and a more effective effort to contain the perceived threats of urban nationalist protest and the rural insurgency. (As is now known, however, Marcos had deliberately exaggerated the extent of the threats to the government from both.)

## Cronies and Technocrats

Previous to martial law, and corresponding to the "elite democracy" then prevailing, the influence or participation of mainstream businessmen in economic policy-making had been generally direct and extensive (Montes, 1988). However, reflecting the political exclusion of the traditional clans and the centralization of authority, economic policy-making under the Marcos regime was concentrated in a small group composed of the Marcoses and Romualdezes themselves, the cronies, and the so-called "technocrats," and to an extent, multilateral institutions such as the IMF and the World Bank.

Besides cooption and expropriation, Marcos had attempted to supplant the hitherto dominant traditional clans with a new set of business leaders absolutely loyal to himself, the so-called "crony capitalists". Some of these had been prominent before martial law. Others were simply beneficiaries of decreed wealth and fictitious capital. Whether old-money or new-money, however, the cronies benefited through the assignment of privileges by the government, first through the deeding over of the property formerly owned by the oligarchs (e.g., broadcasting and newspaper facilities, public utilities, agricultural lands, financial institutions), the assignment of trading monopolies and an array of other exclusive privileges, commissions, and guarantees (de Dios, 1984). The expanding role of government under the dictatorship greatly facilitated the process. An important manner in which crony businesses profited from their government connection was in the matter of government guarantees on foreign loans (Lind, 1984). Almost all crony projects were highly leveraged. However either because of corruption, incompetence, and/or adverse external conditions, many of these ventures failed, leading to the assumption of their debts by the government. The financing of crony-related investments and projects constitutes a large explanation for the magnitude of the external debt accumulated during the Marcos regime.

The fact that some traditionally prominent clans were persecuted by Marcos while crony capital was fostered, initially encouraged a theory—later disproven by events—that the cronies represented a sharp break from the development path followed by the old elite. However, it is evident the crony phenomenon was no more than a logical extension and culmination of the premartial law process of using access to the political machinery to accumulate wealth.[3]

It should be noted, however, that outside of a few, the dictatorship kept cronies from assuming formal positions in politics or in the bureaucracy. The privilege of combining both business and political influence was reserved for members of the Marcos and Romualdez families themselves. Instead, in a deviation from previous practice, the regime began appointing technocrats to Cabinet positions and other important offices.

The technocrats provided a counterpoint to the cronies—though ultimately not enough to be counterweight. Principally a Marcos innovation for the Philippines, (in fact antedating the imposition of martial law), the technocrats were persons appointed to high government positions on the basis of their academic or professional credentials. They were put in charge of the Central Bank, the Ministries of Finance, Industry, Agriculture and Planning among others. This represented a change from the pre-Marcos period when prominent business people participated directly in policy-making. What made the technocrats a curious social phenomenon was that they had no obvious domestic constituency (aside perhaps from Marcos himself) and seemed more prone to represent the views of international lending institutions (which, to be fair, were no doubt also their own views). But for these very reasons the technocrats were a serviceable element in the Marcos dictatorship. Having no stable domestic constituency,[4] they were ultimately beholden to Marcos for their positions and therefore vulnerable to being overriden in their decisions, as they were on many occasions. On the other hand, their training and world-view made them the ideal channel for justifying the continued access to financing given the regime by the IMF and the World Bank.

This was true despite the fact that the technocrats subjectively regarded their connections with international institutions as a form of leverage to introduce reforms in the regime. As G. Sicat from the Ministry of Planning is said to have commented, "We are using the World Bank to add gravy where we need it." (quoted in Broad [1988]). But the situation was rather like that of a bird riding a bullock, believing it steered the great beast. In the course of the dictatorship, however, it would become evident that the technocrats carried no real weight in the government and that the agenda was ultimately set by business and political interests closer to the Palace.

In the last days of the regime, owing to their conflicts with the cronies, some technocrats (especially the Prime Minister C. Virata) had gained some support from the non-crony business sector, but most of them remained personally loyal to Marcos. Despite the aura of professionality surrounding them, their function had objectively been to legitimize the dictatorship and its policies to foreign creditors and ensure continuing access to capital markets for the dictatorship's projects. It was under a technocratic central banker and a finance minister that, in 1983, the government deliberately overstated the country's international reserves. The record of technocrats in the government under Marcos has gone a long way towards reducing the prestige of academics in government.

## Nationalists and Transnationalists

From the 1960s onwards, the outstanding economic agenda had been one of search for a replacement to the import-substitution (IS) regime and the solution of the economy's chronic tendency to crisis and current account deficits. The alternative then (as now) being pushed by multilateral institutions such as the World Bank and more recently the IMF, was the so-called "outward looking" strategy, or export-oriented industrialization (EOI), in ostensible emulation of the experience of the new industrialized countries. This called for the removal of exchange-rate restrictions and quotas, and the lowering of the average level and spread of tariff protection, or in short, a freer or more neutral trading regime. Other observers, however, (e.g., Bello et al., 1982 and Broad, 1988) have regarded this as an attempt by transnational interests to foist a patently unworkable model of development on developing countries to fix them to an international division of labour favourable to the developed countries. In the Philippines, export-oriented development has been represented mostly by the technocrats during the Marcos regime and had a tradition within the planning ministry NEDA (National Economic and Development Authority). The appointment of people who seemed to share a vision of trade and market-oriented development encouraged the multilaterals to continue supporting the Marcos regime (Thompson and Slayton, 1985).

As it turned out, however, the issue of the development strategy could be essentially avoided throughout most of the Marcos regime. Up to 1980, huge amounts of funds could be raised from private capital markets abroad and the current account deficits could be financed without adjustment; hence the pressure from the multilaterals to adopt the EOI strategy could be effectively deflected. Only around 1980, when international conditions had become much less favourable and the dictatorship was becoming hard pressed for funds, did it sign a structural adjustment loan with the World Bank to finance a trade-reform program (de Dios, 1988 and Fabella, 1989).

## Nature of Elite Interests

The pattern of Philippine development experience has more or less conformed with the sequence of trade and industrialization strategies typically undergone by other middle-income developing countries. Up to the late 1940s, the country's trade pattern had been typically colonial, that is mineral and agricultural exports in exchange for imported manufactures. The era of import-substitution began in 1948, when as a result of a large payments imbalance, import quotas were imposed. The import-substituting period also spawned the Filipino manufacturing class. In current discussions regarding trade policy, some ideological confusion is caused when the import substituting sector is depicted as one with a distinct nationalist orientation. What is crucial to note is that it was many of the landed gentry themselves who diversified into import-substituting industries, even during the early era of protectionism, without totally relinquishing landed interests (Carroll, 1960). In an important sense, therefore, unlike the classic development of capital, the manufacturing interests did not represent a readily distinct social class from the aristocracy.[5] Secondly, evidence suggests (de Dios, 1986) that beneficiaries of protection were firms with an appreciable degree of foreign ownership, such as tariff-jumping subsidiaries or foreign licensees who were dependent on access to foreign technology and inputs and the foreign exchange to pay for them. Historically, the dependence of the manufacturing sector on foreign exchange has never been overcome, which is significant in explaining the conservatism of the elite with respect to the stance on the debt issue. In any event, by the early 1960s, immediately before the Marcos presidency, the limits of the *easy* phase of import substitution had been reached and the country signed its first IMF standby agreement leading to a peso devaluation. By then the structure of trade had changed so that imports consisted mostly of capital and intermediate goods, rather than consumer products; the latter now being produced domestically.

A crucial factor preventing further industrialization along the import substituting route was the narrow base of incomes constituting the domestic market. Even as protection to inward-looking industries continued by means of quotas, multiple exchange rates and tariffs, few industries were able to attain scales large enough to permit backward import-substitution or export competitiveness. The narrow market was traceable to the highly skewed distribution of assets, particularly agricultural land. But, precisely because of the virtual identity between manufacturing and landholding interests, there was unlikely to be any abiding interest among the elite to support redistributive efforts such as agrarian reform programs. On the other hand, one always expect a large constituency for increased government spending, regardless of fiscal consequences, as it would stimulate domestic demand without affecting the distribution of property. A consequence of this, however, is the tendency for the economy to run up large external deficits and run into difficulty with creditors.

This observation is important for an understanding of the direction of economic policy-making in view of the earlier discussion of patronage politics. In spite of its centralizing tendency and apparent bureaucratizing elements, martial law in the Philippines did not (perhaps could not) override the particularism of interests which were the staple of premartial law politics. Marcos reaffirmed the symbiotic relationship between local politicians and the central government. Elections for local officials were provided for even within the dictatorial framework and were allowed to reflect the balance of economic and political power in the localities. Local personalities, however, were in turn expected to be loyal and deferent to the central government (the dictatorship) in *larger* decisions and, in turn, benefit from public works and other transfers.

Summarizing the changes introduced by the advent of martial rule with respect to the actors in the making of economic policy, would include: (a) The replacement of politically prominent businessmen by cronies assigned to specific sectors of the economy. The most prominent examples of these were the economically strategic coconut and sugar industries; (b) The crony system was augmented by technocrats who served as a link to international financial institutions at a time when these were crucial to financing deficits; and (c) The patronage system at the local level was basically retained, with a dependent relationship fostered between the particularist interests of local politicians on the one hand and the highly centralized national government which dispensed favours on the other.

Politically, these changes were reflected in the abolition of the old Congress through the imposition of martial law. Between 1972 and 1980, the main instruments of legislation were executive acts, such as presidential decrees and letters of instruction. Influence in decision-making, both political and economic, was measured by one's access to the President's Study Room (where Marcos signed documents). The convening of a rubber-stamp assembly in 1978, in principle supplemented the president's decree-making power with that arising from the legislative. But the initiative for legislation remained with the ministries and offices associated with the executive formally under the rule of the technocrats. However, even the technocrats were only allowed to "formulate and rhetoricize the public agenda in the form of economic and development plans which formed the basis for foreign loans. The political leadership then allowed the unconstrained introduction of exceptions that made a complete mockery of the spirit and letter of the plans" (Fabella, 1988).

As it pertains to the matter of the foreign debt, the centralization of decision-making power was mirrored in the absolute authority vested in the president to contract foreign loans, without legislative review. It could also be seen in the unlimited discretion exercised by the president in picking the investment projects to be financed by large government financial institutions such as the Philippine National Bank (PNB) and the Development Bank of the Philippines (DBP). Many of what are called "loans upon behest" (a good number being facilitated by no more than marginal notes written by Marcos) turned out to be bad investments and made up the bulk of the nonperforming assets held by such financial institutions after the crisis struck.

The acceptance of the dictatorship depended upon its ability to sustain the interests of the rest of the business and landowning classes; those which were not directly involved in the immediately preceding political conflict. In this respect, the world economy cooperated with the regime's need to consolidate itself. High commodity prices in the early 1970s cushioned the impact of the oil-price shock and, in the second half of the 1970s, easy foreign commercial credit became available. (The country had all this time been almost constantly been under IMF supervision either under some standby credit arrangement, a source of reassurance to the foreign commercial banks.) The combination of good export prices and low real interest rates represented the best of all worlds for the Marcos regime and allowed it to consolidate its hold on domestic politics through expanded government or government-guaranteed spending, most of it using foreign finance. Philippine external debt rose from $2.7 billion in 1972, to $10.7 billion in 1978 and $20.9 billion by 1980. The good times ended with the 1980-81 recession in the developed countries, which brought both lower export demand and higher interest rates. Between 1980-1982 the government chose to pursue a countercyclical spending policy, hoping to *ride out* the recession. But reserves dwindled even further (despite their overstatement) and were sustained only by continuous short-term borrowings and outright overstatement, the magnitude of which would only later come to light. In August 1983, Benigno Aquino was assassinated causing unprecedented protest and political uncertainty; short-term creditors did not renew their loans, and the Philippines declared a moratorium on debt-payments in October of the same year. The subsequent withdrawal of support from the dictatorship by the business community during this period contributed materially to the subsequent downfall of the Marcos regime.

## 3. The Constituency and Agenda of the Aquino Regime

In order to address the reasons for the continuity in certain economic policies under the Aquino regime we must briefly inquire into the nature of the circumstances and the coalition that brought Mrs. Aquino to power.

### *Resurgence of the Crisis*

Until about 1981, despite the depredations of crony capital and an expansionary fiscal policy, the dictatorship had succeeded in placating, if not coopting, the rest of the non-crony elite without having to substantially alter either property relations or the structure of economic incentives. This was possible as the availability of foreign credit without conditionality and the continuing support shown by multilateral institutions had allowed the deficits to be financed, where under other circumstances they would have led to a payments crisis or speculative attacks on the currency. With the recession abroad, the rise in interest rates, and the start of the debt crisis with Mexico, severe payments difficulties arose for the heavily leveraged crony firms. These firms, together with their debts, were taken over by the government which then had to subsidize them from budget. The selectivity in the "bailouts" and the privileges given to the previous owners were a cause for resentment among the non-crony business sector, most vocally represented then by Jaime Ongpin, later appointed and dismissed as finance minister under Aquino. Business-sector criticism of the dictatorship and crony capital mounted—alloyed with indignation over the Aquino assasination—especially after the payments moratorium and the suspension of normal relations with foreign commercial banks. Therefore, in the midst of economic crisis there had occurred another split in the elite.

Two consequences of the 1983 debt crisis were especially distressing to the business sector: the foreign-exchange rationing during the period of the moratorium; and the deep recession that began in 1984 as a result of the stringent monetary targets imposed by the IMF under a standby program. By this time, after the discovery of the reserves overstatement, the IMF had "changed its posture from that of a doting parent to that of a vengeful god" (Montes, 1988:15). The interests of the import-dependent and domestically oriented business sector were very sensitive to the denial of access to foreign exchange and a downturn in domestic demand. These two events determined the business opposition to the Marcos regime.

In many important respects the coalition that supported Corazon Aquino in the elections and took part in the February Revolution, sought to represent large business and landowning interests, whose main demand was a return to premartial law order and property rights and to normal business conditions. The greater part of dissatisfaction with Marcos among the businessmen was the latter's inability to command credibility among the international banking community. This lack of confidence argument was often heard at the time, with the implicit assumption made that political legitimacy may ultimately be equated with support from business and banking. The most prominent representative of this point of view was the late former Finance

Minister, Jaime Ongpin. It was he, together with J. Fernandez, who negotiated with the banks for the new Aquino regime and came up with the present restructuring agreement, for 1988-1992.

## Least Common Denominator

The complication in this otherwise neat story, however, is that it was not simply the ruling classes which had participated in the election and revolution—otherwise there would have been no revolution. In order to succeed, the Aquino coalition had to make common cause or, at the least, not directly antagonize anti-Marcos groups with more populist or radical demands. The programs of the Aquino coalition had to tread carefully between the interests of the masses and the elite, reducing the agenda to some least common anti-Marcos denominator. In the "economics" speech before businessmen in 1986 (reprinted in Daroy et al., 1988:694), the stress was on topics not controversial within the Aquino camp such as a reduction of unemployment, a renegotiation of the debt, dismantling of the trading monopolies and withdrawal of crony privileges, encouragement to small-scale industries, and reduced government intervention. Much of the intellectual spadework for this had been provided by an independent group of academics from the University of the Philippines School of Economics, who in 1984, published a report (de Dios, 1984) criticizing the role of the dictatorship's policies in the outbreak of the economic crisis.

In the period immediately after the new regime came to power, before the bureaucracy had recovered and political relationships ossified, a large role could still be played by less traditional forces within the Aquino coalition. This was reflected in some Cabinet assignments and in the appointment of "officers-in-charge" to (formerly elective and elite-dominated) local government positions.

Owing to the its early receptivity to nontraditional elements, the new regime's economic plans could be influenced by the collective efforts of researchers and academics (encouraged by the newly appointed planning minister, S. Monsod) in two reports, "Towards Recovery and Sustainable Growth" (Canlas et al., 1986), and "Economic Recovery and Long-Run Growth: An Agenda for Reforms" (Alburo et al., 1986). Although these were documents of compromise and consensus among their writers, they were distinguished from earlier ones by their espousal of two measures which were controversial among the businessmen and landowners, namely, selective debt repudiation and wealth redistribution, to be achieved mainly through agrarian reform. Their basic insight was that recovery would be hampered by the outward resource-transfers represented by debt repayments. Hence the government was urged to make use of its moral position and international prestige to limit debt payments to what was consistent with rapid growth. Secondly, while supporting the dominant development wisdom about the need to "get prices right" (especially prices of factors and tradeable goods) the reports argued for a thoroughgoing redistribution of wealth, in particular through taxation and agrarian reform, and through these measures, the creation of a more buoyant domestic market. The second report was markedly less sanguine about the sufficiency of anchoring development solely on the growth of world trade, or the sufficiency of export-orientation.

By the time these ideas had filtered through to the government's own document, however, they had lost their cutting edge. For although the NEDA staff may propose plans and strategies, adoption rests with the Cabinet and NEDA board, headed by the president herself and consisting of several Cabinet secretaries concerned with economic and social matters. The NEDA secretary-general (who is ex-officio planning secretary) merely heads a staff and sits only as one among many members on the board.

In the "Agenda for People-Powered Development" (1987), the document finally adopted as the government's recovery plan, the issues were framed in the less controversial terms of recovery of per capita real incomes, to be achieved mainly through increased spending on government compensation and infrastructure, privatization of crony capital and market deregulation. In effect, the ideological self-differentiation of the Aquino regime had been halted with the limited agenda that the government set for itself, especially during the early years. In turn, this limited agenda may be explained by the need to maintain a broad anti-fascist alliance, and possibly traced to the ad hoc nature of the Aquino candidacy itself. In a sense, the heterogeneity of forces which allowed Mrs.

Aquino more easily to come to power also made the consolidation of this power more difficult.

## Drift to Conservatism

During its three-odd years in power, different forces, circumstances, as well as its own choices, have pushed the Aquino government gradually towards a more restorationist or conservative path. One reason is the legitimate representation within the Aquino coalition of elite interests who opposed Marcos, including large landowners and businessmen in manufacturing and finance, who now felt justified in laying claim on their share of the economic and political opportunities provided by the revolution. An important indicator of this was the appointment of prominent businessmen to important government posts, a process continuing up to now. In contrast to the Marcos period when technocrats were placed in charge of similar ministries, prominent pro-Aquino businessmen are now to be found heading the departments of finance, trade and industry, the Central Bank, and many large government corporations. In this respect, the Aquino government was doing nothing more than restoring a prevalent practice of premartial law politics. At present the only cabinet level appointee with an exclusively academic background is the secretary of planning. But a no less important reason has been the political need to heal the split within the ruling elite, this time between the anti-Marcos factions and those who had collaborated closely with the dictatorship. Concretely, this process of reintegration has taken several forms. Some important officials from the Marcos regime have simply been retained. Among the most prominent are the secretary of defense, Fidel Ramos, and the governor of the central bank, Jose B. Fernandez, a Marcos appointee, but related by affinity and association to the Cojuangcos. While it is true that in some cases estrangement followed, this occurred only after the agenda for the regime had been predetermined in consideration of the participation of the conservative elements. (In the same manner, the successive attempts at military coups have had the effect of diluting the Aquino presidency's initial commitment to the prosecution of human-rights violators in the military and nudging it toward a mailed-fist solution to the insurgency.)

Secondly, through the convening of a Congress, an arena was opened for legitimizing political participation of traditional politicians and members of the elite, including former Marcos loyalists, who by no means supported the February Uprising. Unfortunately, the *new* Philippine Congress has a distinctly musty smell since, by one count, roughly 65 percent of it is composed of traditional politicians and 35 percent from clans associated with the Marcos regime (Soriano, 1987). The local elections have fulfilled a similar legitimizing function (in one case permitting a soldier-warlord suspected of human rights violations to become a governor of a province). This must be contrasted with the demand of cause-oriented groups, turned down by Mrs. Aquino, that she use the revolutionary power to pass redistributive measures such as agrarian reform, rather than pass on the initiative to a landlord-dominated legislature. By devolving power upon a conservative legislature, the executive power relinquished its prerogative to enact a more thoroughgoing agenda for reform.

Perhaps the penultimate step in the political integration of the elite and the reinstitution of traditional politics was the formation this year of an "umbrella" pro-administration party called *Lakas ng Demokratikong Pilipino* (LDP) among members of Congress. Until then, there had been no single Aquino party (since Mrs. Aquino herself has no party affiliation). The only qualification for joining this new party is professed "support for the administration's programs". Owing to the self-limitations and vagueness of the administration's agenda, however, it is highly unlikely that the new party's platform conceals any more profound principle than that of winning elections. As might be expected, it now commands the absolute majority in Congress, particularly in the lower chamber.

The upshot of this discussion is simply to point out that it will become less likely to see further innovative schemes, whether in economics or politics, from the Aquino administration.

There is one subtle point, however. Previous to the promulgation of a new constitution and the holding of elections for the legislature, the coalition behind Mrs. Marcos had ruled as a revolutionary government and had encompassed heterogeneous forces. A number of liberal and so-called "left-leaning" personalities ran for Congress and, although some won, the majority lost as expected. The effect, however, was to "homogenize" the composition of the executive, which, as an institution, is now much less likely to be adventurous in policy-making than the legislature. One indication of this is the apparently firmer stance the Senate and House are taking with regard to the foreign debt. Bills are now pending in both houses which would compel the administration to lower the net outflow from the debt by limiting the foreign debt-service to a certain proportion of export receipts, an idea the cabinet has balked at.

## Influence of Multilateral Institutions

Another source of continuity is the economic policy advice provided to the government by official multilateral and bilateral institutions, most notably the International Monetary Fund, the World Bank, Asian Development Bank (ADB), U.S. Agency for International Development (USAID), and the Overseas Economic Cooperation Fund (OECF). The prominence of this relationship, however, stems from the fact of the country's indebtedness. With the drying up of commercial sources of capital, the government has been relying mainly on official credit sources. Increasingly, however, credit from these sources has been bound up with conditionality and performance criteria.

The difference between the position of the Marcos and Aquino regimes in negotiating with the multilaterals is that the latter now is faced with less leeway in fulfilling the conditionality associated with official lending, although immediately after the Aquino government came to power and negotiated another standby loan with the IMF, there was actually some relaxation in the targets.

In 1988 the country continued to climb out of the recession and posted a growth rate of 6.7 percent. In 1987 the economy grew some 5.05 percent following a period when real incomes shrunk. As a result of the recent rebound, even the rate of open unemployment has slightly declined, from 11.8 percent in 1986, to 11.3 percent in 1987, and to approximately 10 percent in 1988.

Though providing sufficient political capital, the importance of the recent growth is easy to exaggerate. It has been largely due to reviving consumer demand; real consumption per capita was actually declining for the entire period of 1983-88. In 1988, consumption expenditure grew by 5.57 percent. Construction (mostly in real estate development) has also risen substantially (16.4 percent growth). Nonetheless, even after this latest growth, real per capita incomes are equivalent only to 1976 levels.

A major contribution to the revival of consumption were substantial increases in government expenditure, which took the form of pay-raises to government employees and spending on infrastructure. The large government expenditures made themselves felt in widening budget deficits owing to a perennially low tax effort. As a result, the government has been borrowing heavily on the domestic market (T-bill rates rose to some 18 percent in 1988) as well as relying on foreign financing. The present rapid growth raises the question whether it can be sustained or will soon hit an external-balance ceiling. For this reason, the resolution of the debt problem has assumed strategic importance for the administration.

In April, 1989 the government submitted a "letter of intent" or "memorandum on economic policy" (LOI-MEP) towards obtaining IMF loans totalling $1.3 billion over the next two years. The LOI-MEP has been highly controversial owing to the high degree of conditionality attached to it. The government negotiators, however, regarded the IMF deal as a crucial first step towards convincing the consortium of commercial creditors and the bilateral creditors of the Paris Club to lend *fresh money* to the country to reschedule maturing debt. The next section discusses the evolution of the government's debt strategy and the mounting pressures to change it.

## 4. The Debt Controversy

The expansive fiscal policy implemented by the administration in order to sustain economic recovery put it on a collision course with potential creditors, most of whom are reluctant to consider greater exposure to the country. In particular the IMF team (presumably anticipating the objections of the commercial creditors) early on in the course of negotiations with the government, questioned the NEDA staff growth projections, pointing to a public investment bill considered too large (by P17.4 billion) to finance.[6]

The progress of the current negotiations with the IMF and the commercial banks is interesting for pointing up once more a long-standing policy debate between NEDA on the one hand, and the Central Bank and Ministry of Finance on the other. The convening of Congress in 1987 has added yet another venue and set of influences acting on the controversy.

The level of Philippine external debt stood at $28.65 billion as of March 1989 or some 83.8 percent of GNP. Debt-service amounted to $3.0 billion annually, representing 33.2 percent of exports in 1987. Similarly 32 percent of the national budget went to servicing external obligations for 1988. (To the extent the loans were not contracted by the Aquino government, it is the debt overhang which causes the deficits, and not the other way around).

The first dealings the new Aquino government had with the commercial bank consortium came in March 1987 when a rescheduling of commercial loans was due. The negotiating team appointed by Aquino consisted of Finance Minister Jaime Ongpin and Central Bank governor Jose Fernandez, with the NEDA's Monsod playing a subsidiary role. (The indispensability of the NEDA head stems from the macro-economic and planning capability of her staff.) Fernandez himself was no new face, having negotiated with the banks under Marcos as well. Ongpin and Fernandez negotiated a multi-year rescheduling, albeit with interest margins higher than on similar earlier deals made by Argentina and Mexico.

Reportedly against NEDA's advice, no "new money" was requested.[7] The agreement came in for a good share of criticism among media and academics. Under these arrangements, the expected net resource outflow from the country to lenders for the period 1988-1992 would be in the order of $15 billion, a form of reverse mini-Marshall plan from South to North. With the rapid growth of expenditures last year, however, even the Central Bank and Ministry of Finance have belatedly realized the necessity to approach the commercial banks for new or fresh money.

## Disputes Within the Executive

Earlier in the presidency, the option of selective debt repudiation was broached most vocally within the government by NEDA's

Monsod. Monsod was one of a group of academics who had been vocal under the Marcos regime and who believed the heavy debt-service was the key obstacle to economic recovery. The notion of selective repudiation itself was already regarded as an effort to moderate demands for a unilateral default or cap on debt-service, which were coming from the cause-oriented mass organizations (the "other" participants in the EDSA uprising). The most articulate proponents of an assertive debt strategy of outside government were grouped in the Freedom from Debt Coalition (FDC). The FDC was proposing a change in the negotiating stance of the government, a limit on debt service equal to 10 percent of export earnings, and the complete repudiation of fraudulent loans.

In theory selective repudiation meant the new government would not assume obligations which benefitted only Marcos cronies and the foreign commercial banks. A prime candidate was the Bataan Nuclear Power Plant, whose construction by Westinghouse and financing by the U.S. Eximbank was strongly suspected of having been attended by corruption.[8] In the Cabinet, however, Monsod's position was opposed by then Finance Minister J. Ongpin and the Central Bank Governor J.B. Fernandez, who warned of cross defaults and the cutting off of international credit. Ongpin, a former chief executive of a transnational mining and construction company, and Fernandez,[9] erstwhile head of a large commercial bank, were echoing the anxieties of big business and finance. It did not help either that in Cabinet meetings, in which the Central Bank governor also sat, discussion of the debt and financial matters either did not come up, or went over the heads of other ministers, tending rather to be dominated by Fernandez and Ongpin.

NEDA's plan to sustain the growth momentum by deficit spending (the quaint Keynesian term "pump-priming" was used) ran into conflict with the passive stance the government had taken with respect to the foreign debt. On the one hand, short of a large-scale asset redistribution, the government would have to rely on its spending (as in the past) to stimulate domestic demand. But without a huge tax effort (also redistributive), this would entail either an increase in foreign financing or at least a cut in outward payments. Short of some form of a debt-cap, therefore, the continuation of the recovery hinged upon the securing of new financing from the commercial banks and official creditors. (The

NEDA estimate of the financing gap for 1988-1992 is $7.7 billion.) For NEDA the main goal was to support a higher rate of growth, targetted at some 6.5 percent annually. Whether intended or not (and no longer arguing about its economic soundness) this goal was also politic for a government confronting threats to its legitimacy.[10] It was ultimately for this reason NEDA recommended debt-relief through selective repudiation. Of course new financing was another alternative but, in the first place, it was held unlikely that the foreign banks would increase their exposure to the country. Second, negotiating with commercial creditors presupposed a parallel deal with the IMF, which was likely to place a lid on investment and growth targets. But, in any event, NEDA's was willing (albeit skeptically) to take the new money position as a second-best alternative to a cap on debt service, so long as the rate of growth (with its implied levels of public investment) was not compromised.

As already mentioned, no new-money requests were in fact made by the Fernandez-Ongpin panel in the March 1987 rescheduling talks with the bank consortium; the argument made was it was not needed at the time. This led to a minor dispute with NEDA, which was projecting a large financing gap owing to an anticipated economic expansion. Subsequently, internal politics and horse-trading within the executive, as well as widespread criticism of the disappointing terms obtained by the debt-restructuring talks, finally led to the replacement of Ongpin by Jayme in a Cabinet shake-up; inexplicably, both protagonists Fernandez and Monsod were retained.

As the economy began to pull out of recession, however, it became obvious even to the Ministry of Finance and the Central Bank that a large financing gap for 1988-92 would have to be filled. (An undeniable indicator was the continuing huge drawdowns in international reserves, which fell from 3.2 months worth of imports in 1986 to 2.1 months of imports in 1988). By late 1988, the Fernandez-Jayme panel had also adopted the new money line. This time the position of the Central Bank and Ministry of Finance seemed to be that, considering the urgency of the situation, the first priority was to obtain an agreement with the IMF, and ultimately the commercial banks, at all costs and as soon as possible. NEDA's concerns about the rate of growth and the public investment level were regarded as mere quibbling. Most prominent business-sector elements were also against any suggestion that the country ought to take a harder line with the creditors.

From time to time, the president's prepared public speeches contained a hint of *toughness* on the debt issue. This played to the crowds, since, especially during the earlier pre-legislature period, there was an expectation Aquino could use her obvious reservoir of public support and international goodwill to obtain a better deal from the commercial banks.

Ultimately, however, despite occasionally brave rhetoric, the official position settled down to one of honouring all obligations, including loans incurred by the previous regime. By mid-1988, the meaning of the "growth-oriented debt management strategy"— ostensibly the government's *new* approach to the debt problem— had narrowed down to seeking to reduce the net resource outflows by negotiating with the commercial banks for new money and with the Paris Club countries for greater official development assistance (ODA). In order to achieve both, the executive considered it indispensable to conclude a standby agreement with the IMF for the "seal of good house-keeping".

## *Letter of Intent*

As expected, the initial consultations with the IMF technical staff regarding the emergency fund facility (EFF) and the drafting of the government's letter of intent did not go smoothly. The government encountered difficulty negotiating another standby credit with the IMF owing to disagreement over the achievable level of the public deficit and the public investment targets. The difficulty lay in the initial presumptions from which both sides started. NEDA perceived its minimum mandate as the achievement of recovery in real per capita incomes by the early 1990s. Hence its existing plans (which were after all officially adopted documents and formed the initial basis for discussion) proceeded from desired growth rates, to investment levels, and arrived at foreign financing requirements, given trends of savings, exports, and tax-revenues. The IMF staff tended to proceed on what was the likely expected level of foreign financing in anticipation of the preferences of the foreign bank consortium.

From a sense of anxiousness to conclude the negotiations, but perhaps also owing to close dealings with the banking community, the finance-central bank staff tended to internalize the IMF's own position. From this viewpoint the NEDA growth figure of 6.5 percent and corresponding public investment targets were indeed "too high". By this time, however, public opinion had been accustomed to thinking of recovery as a sine qua non; 6.5 per annum had become a kind of icon, and a reduction of the target growth rate would have been regarded as unpopular.

The matter was finally resolved by virtually excluding NEDA from participating in the final negotiations with the IMF and the drafting of the LOI. In a confidential letter to the president (later leaked by media) regarding the LOI, Monsod reveals that "NEDA's participation in the four-week negotiation with the IMF team was partial and limited; except for one internal policy level discussion which I attended, NEDA was not involved at all in the policy level discussions nor in the drafting of the program."[11] The final version of the LOI retained the NEDA's growth target of 6.5 percent but fudged the figures on how to achieve it. High income growth was made consistent with lower ceilings on budget and current-account deficits simply by projecting higher than usual export growth, unusual tax collection efficiency, and lower deficits for government corporations.[12] An important means for achieving the latter was increasing the prices of rice and rates for water and power.

This device merely resolved on paper the conflict between growth and debt service. If the IMF were to insist on the ceilings on the current account and budget deficits, the only way to meet them in the face of actual exports and tax collections would be to reduce public consumption and investment. This by itself, however, would reduce the rate of growth. On the other hand, with those ceilings, the estimated financing gap would be much smaller and banks would not need to lend as much.

Apart from technical objections to it, some features of the LOI struck a sensitive nerve among the rest of the population which transformed it from an esoteric concern to a poor man's issue. The impending reduction of the subsidy to rice unleashed a wave of speculation and led to immediate price hikes. Other implied price increases for utilities in the LOI compelled informed unions to revise their expected inflation estimates as well as their minimum-wage demands. This placed them on a collision course with employers' associations and the executive (as well as NEDA) which were willing to accept only a smaller increase in the minimum wage. (As of this writing, organized labour is threatening work stoppages if a minimum-wage increase of P30 is not granted, while the president has left the issue to Congress.) Reflecting this concern, the activities of the FDC have revived with the launching of a specific Campaign Against the Letter of Intent (CALOI) in which workers and urban poor play a more pronounced role than before.

In the end, debate over the IMF letter of intent within the executive terminated when it was accepted by Mrs. Aquino as being "the best available under the circumstances". NEDA's Monsod was chided and tolerated in the cabinet only if she kept her opinions to herself.

From a wider perspective, the alignment of forces on the debt issue is hardly surprising. The conflict within the cabinet may be regarded as part of the process by which conservative elite elements seek to steer the administration along less adventurous and unorthodox lines and consolidate their influence on policy. An important reason the big business sector rejected the Marcos regime was its loss of credibility with international creditors and consequent inability to supply the needs of a dependent economy. The adoption of any independent debt initiatives would certainly be disruptive of interests in existing trade and capital flows and counter to the overarching need of the elite to maintain its links with transnational finance.

The triumph of the finance-central bank line on the debt issue owes mainly to political-economy factors relating to the changing nature of interests represented within the executive. But reasons relating to mechanisms and processes of decision-making are also important. For example, there is only a limited flow of economic information at the cabinet level, as well as limited capacity to absorb it. This prevents the evolution of a meaningful consensus on issues and accords a dominant position to the central bank governor and finance secretary providing, of course, they have the president's consent. Efforts of other cabinet members to inform themselves are reportedly discouraged (indeed, at times positively suppressed) and viewed as undermining the authority of the debt panel. Even during the negotiations with the IMF technical staff, the NEDA staff contribution was excluded. The entire episode reflects a decision on the part of Mrs. Aquino to continue to trust her negotiators regardless of contrary views. It is in this fashion that an executive stance has been held up and the focus of the debate then shifted to Congress.

## *Congressional Initiatives*

If it has found no audience in the executive, the debt-cap seems to have found at least some support from members of the legislature, especially in the Senate. A senate bill and at least two pending house bills have the effect of limiting debt service payments to a level considered tolerable for the country (e.g., at most 20 percent of total exports in the senate bill and 10 percent in the house version). Other initiatives would reduce the executive branch's prerogatives in conducting debt policy and vest additional powers of review with Congress. Among these bills are those which would (a) reserve for Congress the right to review foreign loan contracts entered into by the executive; (b) repeal a provision which provides automatic budgetary appropriations for servicing past debts; and (c) set up a joint executive-legislative debt council (JELDC) to set policy on foreign debt.

The rationale for many of these bills is principally a genuine dissatisfaction with the manner the executive branch has conducted debt policy. But, in addition, they are also attempts to assert the influence of Congress over what it regards as its domain. A good number of legislators believe their budgetary powers are severely infringed upon as a result of the large chunk of the annual budget automatically appropriated for debt-service. Some 27 percent of annual appropriations is currently devoted to foreign debt service. Finally, of course, there is simply the fact that the debt issue is a popular one with visible and vocal constituencies. For example, intensive lobbying and public information campaigns by FDC-CALOI have played a significant role in nudging these bills forward in Congress. The readiness with which some legislators have embraced a popular cause was influenced in no small way by the need to revive the reputations of both chambers of Congress after recent scandals involving imports of firearms and cars for its members.

Interestingly enough, the dispute between the President and Congress over debt policy has taken the somewhat superficial form of delimitation of functions. The issue finally came to a head when both houses of Congress overwhelmingly passed a bill in February 1989 creating a Joint Executive-Legislative Debt Council (JELDC). With members to be appointed both by Congress and the executive, its function was to actively conduct debt policy. While the strategy to be pursued by the proposed debt council was not specified, it was supposed to seriously consider hitherto controversial measures such as a debt-service cap and repudiation of all or part of the debt, among others options. Arguing that only the executive had the right to conduct debt policy, Mrs. Aquino vetoed the bill, the first sign of open conflict between Congress and the executive under the Aquino administration. Considering the vote on the bill, Mrs. Aquino's veto could easily have been overridden. Rather than risk this, Mrs. Aquino broke protocol and met with Congressional leaders in a closed-door session to dissuade them from overriding the veto. At this crucial stage, the administration party LDP (the majority of which had actually voted for the bill) cracked the whip, and the veto was sustained. Ultimately a revised version of the bill was passed into law under which the debt council would merely advise the negotiating panel.

Similarly, after the LOI had been drafted for submission, several senators and congressmen decried the secrecy in the process and demanded that the legislature be consulted and informed regarding the commitments the executive was making to the IMF. The argument was again couched in terms of jurisdiction; budgetary powers belonged to Congress and the contraction of new loans ultimately involved appropriations for their repayment. Copies of the LOI were given to congressmen and its contents became public. When a substantial number of legislators expressed their opposition to the program under the LOI (owing partly to the negative public reactions to it), Mrs. Aquino once more "consulted" with members of congress. Some senators argued that government would not be bound by the terms and conditions of the agreement unless two-thirds of the Senate concurred, but were ultimately persuaded by Mrs. Aquino's argument that the agreement was "the best available to the country under the circumstances". Apart from Mrs. Aquino's vouching personally for the LOI terms, another persuasive argument for Congress was that IMF approval was needed if the planned increase in official aid, known as the Philippine Assistance Plan (PAP), was to materialize. When asked by Mrs. Aquino for its comment on the terms of the submitted LOI, the Senate in the end voted 13-7, with significant absences, to endorse them. On two issues of the debt, therefore, namely the debt council and the LOI, Mrs. Aquino had exacted surrenders from the House and the Senate, respectively.

## Philippine Assistance Plan

The connection between the Philippine Assistance Plan (PAP) and the adopted debt strategy is worth examining in some detail. The PAP's final shape is still unclear, but a fair description of its present status, is that it is a campaign by the Philippine government, encouraged by the U.S., to persuade bilateral donors, especially Japan, to substantially increase their official assistance to the country. For some unknown reason $10 billion over five years has been touted as a likely figure. To those who were arguing for pliancy on the debt issue, PAP was regarded as the compensation for the good behaviour the country shows in the negotiations. It is argued that official development assistance would more than make up for the admittedly harsh conditions imposed by the IMF and the bank consortium in regard to the commercial part of the debt.[13] This argument has been especially disarming to local officials and politicians who are led to believe—in a modern variant of the airplane cargo-cult—that with PAP, the national government will be able to accommodate their favourite public works.

It should be noted, that the PAP machinery is being set up largely outside and independently of the usual government bureaucracy. Its appointed head is a prominent businessmen, R. Villanueva, head of a large construction and development company. PAP is supposed to come up with new projects for the aid pipeline and, although the funding sought is public, suggestions from private business are given particular attention. This may be seen as an effort to short-circuit the usual processing of ODA, which typically passes through NEDA. NEDA, however, is being perceived or depicted (by finance among others) as being "too bureaucratic" and "too slow", a criticism which refers mainly to the formal requirements imposed by NEDA on projects submitted for funding under the official assistance projects. Under existing arrangements, NEDA's project implementation staff requires detailed feasibility studies, a procedure that screens out potentially bad projects but in the view of aid enthusiasts, slows down the country's ability to "absorb" official assistance. Finance officials particularly have publicly blamed NEDA's procedures for the slow disbursement of committed ODA funds.

Deeper reasons, however, may lie in the recalcitrance of NEDA (and in Monsod in particular) in the debt-negotiations, which is incongruous with its prominent role in the disbursement of official assistance. Shutting out NEDA in the public-sector component via PAP, however, conveniently undercuts it in the private debt negotiations as well.

In the rush for ODA, the real possibility is overlooked that the government may indeed have only a limited capacity to generate viable projects in a given time. An important reason ODA cannot completely substitute for debt relief is that while both may go the same way towards filling an aggregate financing gap, the quality of investment choice implied in each may be quite different, since different agents (the government in the one, private entrepreneurs in the other) are making them according to different criteria. The incorporation of private-sector decisions in the use of public loans (as in PAP) does not diminish the potential for poor choice of projects. On the contrary, it increases the scope for rent-seeking.

## *Patronage Politics and National Issues*

The dynamics (or rather the lack of it) of the debt issue may be explained through a more general description of the evolving constituency of the Aquino regime. The government's "revolutionary" agenda and constituency were eroded with the reintegration of estranged elite factions into the political mainstream and the beginning of legislative participation. This effectively shifted the executive's concern away from any radical measures, leaving controversial issues that required positive action such as agrarian reform, trade and industrial policy and more recently, wage policy, to Congress.

The earlier influence by liberals and progressives on policy-making was displaced from the executive branch to the convened Congress, especially the Senate. This shift is most visible in regard to the debt issue, as former Cabinet members with ideas on alternative debt strategies left to run for the Senate. These new elements in the legislature were nonetheless overshadowed by the renaissance of traditional politicians, who ultimately came to dominate legislative agenda-setting.

By the same token, this process also homogenized the composition of the executive nudging it towards greater conservatism. The intellectual or professional component of the executive, hitherto accessible to centre-left influence, could then pass into the hands of more traditional business leaders and bureaucrats, who joined with original conservatives in the Cabinet and holdovers from the Marcos regime. Owing to the previous disrepute of technocrats under the Marcos regime, their appointment, without clear constituencies, was hardly pursued.

At the same time, it would be an oversimplification to say that the entire decision process may be explained as a straightforward implementation of conservative elite interests. The illustrated conflict between the executive and the legislature over the debt issue calls for a more differentiated analysis.

The dynamics of politics on national issues such as the debt is an outcome of a tension and symbiosis between national and local politics, as expressed roughly by the executive and legislative, respectively. Lande's[14] description of premartial law politics, though not exactly using the same assignment, is suggestive in positing a functional interdependence between the needs of politicians concerned primarily with national issues, on the one hand, and local issues on the other. National office-holders find local "gentry-politicians" valuable in that the latter can deliver votes during elections. In turn, local politicians must relate to national politics mainly to ensure a continuing stream of public works and services to their constituencies.

The interests of gentry-politicians, however, owing to their particularist concerns, are not always well-defined on all national issues such as an appropriate debt strategy. For this reason, tension is also possible between the short-term interests of self-projection needed to win elections—and in the process taking popular positions in regard to issues—and, on the other hand, the need to maintain a patronage relationship with the dominant elite faction. What is interesting is that while nominal opposition may take place as part of a bid for publicity, this can never take the form of complete independence from the nationally dominant faction on which particularist interests are dependent. In the debt controversy, this may be seen in the manner the two houses finally gave in to the executive position on the separate issues of the debt council and the letter of intent. The LDP's and Mrs. Aquino's role as enforcers of "loyalty" were also clearcut, the implicit sanction being the withdrawal of Mrs. Aquino's endorsement.

If on the debt issue the centre of initiative has shifted from the executive to Congress, this is because the legislature has become the channel more accessible to popular sentiment (keeping in mind the homogenization of the executive). The Congress, however, is both a channel for conservatism and, to a more limited extent, innovation. That it is equally capable of the former is most directly seen in the emasculation of the agrarian reform code. In the interim the centripetal forces which bind Congress to the president will predominate. Crucial here is the dependence of local politicians on national patronage and the strong direct appeal by Mrs. Aquino to the voting masses. For the moment, this renders politically futile any attempt to openly oppose the president on an issue on which she has staked her personal reputation. The concrete manifestation of this is the organization of the LDP. The cracking of the whip by the LDP is what prevented the confrontation between the legislative and the executive from deteriorating. The senators and representatives were ultimately convinced to acquiesce not by force of argument but by sheer political necessity. One may seriously doubt, however, whether a coalition based not on policy consensus but on expediency can long survive.

## Conclusion

Making sense of debates over economic policy-making is compounded by two things: first, the content of the proposals are themselves empirically difficult to verify, even from the viewpoint of economic theory; and second, even assuming the content of various proposals is sorted out, there is the question of matching these with the interests involved.

This paper has attempted to present a model that could explain the various positions taken by actors within and without government using political and economic categories that correspond as much as possible. In what direction has this led us?

First, it is suggested that the elite, for socio-historical reasons, were not sufficiently differentiated to represent fixed economic interests, even between landowners and capitalists. Politically this is reflected in an undifferentiated politics of personalities and geography. In whichever sector large property is found, it is distinguished by its reproduction through exclusion and monopoly (indeed one might say this is the classic characteristic of landed property). Regardless of the type of economic interest to be promoted or preserved, however, access to the state machinery will always command a premium and struggles are bound to develop to obtain that access. The monopolistic fragmentation or "parcellization" of the economy among the elite, and factional competition as a form of wealth accumulation is one explanation for the violence and corruption typical of Philippine politics.

Secondly, it is suggested that despite their particularist interests, there is a common ground among the elite to support policies which increase domestic, especially public, expenditure; to protect against imports of final goods; sustain a stable, if not overvalued, currency; and resist efforts at wealth redistribution. Combined with stagnant investment and disinterest in industrial innovation, these common needs determine the chronic tendency to run external deficits and the dependence on foreign capital. For the same reason, the nationalist-transnationalist distinction, while suggestive, is probably too roughly drawn. The fact is that the present protectionist interests do not have a stake in the development of the national economy as such, but rather in the exemption of their specific fields of business. These are rather in the nature of "directly unproductive activities" such as tariff-seeking, revenue-seeking, or monopoly-seeking.

Third, international institutions seeking to push rationalizing or liberalizing strategies such as export-oriented industrialization are in the queer position of not having a clear constituency with political influence. These strategies lack any socially influential "bearers"; the technocrats as a group have largely become extinct, and the free-trade ideology they espouse is being resisted by the traditional elite (now holding key positions in government) who have been accustomed to privileged, rent-producing positions in the economy. Hence the changes in trade and industrial regimes envisioned by these institutions will likely be achieved only haltingly and as a result of external impositions.

There are certain consequences of elite fragmentation which make it a serious obstacle to development. Well-known in economic theory are the resource-costs of lobbying and political effort in general. These are especially large in a situation where access to state power is regarded as a major means of accumulation, for the stakes, and hence the warranted expenditure of resources, are then correspondingly large. Another consequence is the inability, owing to factionalism, to mobilize capital for large undertakings. Stock and capital markets are notoriously thin. Instead, the norm is closely-held family corporations grouped around a bank. Substituting for stock-markets are the attempts of "groups" to expand their holdings through state privileges and contracts, transnational tie-ups or outright expropriation under the previous regime.

The argument made for the last dictatorship was precisely the alleged possibility of transcending these costs through a centralization of political power. However we saw that regime went only part of the way in suspending elite-fragmentation, since it continued to affirm the property relations which engendered the latter (e.g., it expropriated only the "oligarchs" and replaced them with its own cronies). For this reason, once favourable external conditions no longer sufficed, the elite crisis reappeared in a more violent form. Nonetheless, to judge by the successive putsch attempts by the military, this conservative agenda is not entirely defunct.

Beyond this, however, the present dynamic in the Philippines comes from two sets of influences which either suspend the elite crisis or induce it to come to a head and, in so doing, resolve it. One path to resolution lies in the country's possible insertion into a favourable world economic environment, just as the availability of huge amounts of capital momentarily enabled Marcos to master the situation. Buoyant world trade, massive foreign participation in the economy, and large amounts of foreign assistance could conceivably suspend the elite's fragmentation by "buying them out" of the decision-taking process. An inkling of this is given by the recent large influx of Taiwanese and other foreign investments, as well as the intrusions of multilateral institutions in policy-making. If successful, these could gradually change the direction and quality of wealth-accumulation. On the other hand, protectionism, world recession, and the problem of negative flows from the debt-overhang itself pose obstacles to this outcome.

There is, however, an alternative being posed by "indigenizing" forces, which would resolve the elite fragmentation by abolishing the property relations on which it is founded. In its most radical form, these tendencies are embodied in the National Democratic Front's program. Which one will ultimately prevail is the overriding question for the country in the remainder of this century.

## Notes

1. In recent economic literature, this phenomenon would perhaps be classified as "rent seeking" (Krueger, 1974) or "directly unproductive activities" (Bhagwati, 1974) or perhaps our notion of the scope of the effects of such activities would be much wider. We shall point out, however, that in contrast with the typical view in the rent seeking literature, we do not consider the presence of such activities to be primarily caused by adherence to a particular trade regime. We tend to take a view closer to that of Findlay and Wellisz (1982), which views the level of lobbying and rent-seeking as an endogenous variable.

2. Historically, the change of collaboration has been an important one that has divided the Filipino elite. Immediately after the Second World War, the political crisis revolved around elite clans (chiefly the Laurels) which had served under a Japanese puppet government, as against those which had sided with the Americans. The crisis was resolved by an unpopular presidential pardon of those involved.

3. If one proceeds from the Krueger-Tullock rent-seeking models, the conclusion must be that the consolidation of rent-seeking activity among a few individuals must raise welfare, since the resources spent on rent-seeking are now conserved.

This might seem like an argument for authoritarian rule. On the other hand, the tradition of Bhagwati and Findlay and Wellisz points to the inherently second-best nature of the problem and yields the possibility that interest groups mutually lobbying against one another might actually result in a superior final position than a case where no opposition is permitted. (See also Brecher [1984]).

4. From 1962 up to the present, the country has always been under some form of IMF supervision or another through some standby credit agreement.

5. Of Marx's two ways in which capital evolves, therefore, the Philippines seems to have followed the first way, in which the landowner becomes capitalist. In contrast, what Marx calls the "revolutionary" way is that in which the craftsman accumulates capital.

6. "... They (the IMF mission) consider our growth targets of an average of 6.5 percent real growth in income as too high and unrealistic. This is mainly because the growth target has an underlying public investment program (inclusive of the national government, government corporations, and local government units during the next four years amounting to P247 billion which is bigger than what the IMF considers financiable. On the other hand, the IMF growth scenario has a public investment program that is P17.4 billion lower than the Philippine program... (T)he facts provide the basis for the Philippine program's assumptions, contrary to IMF fears that they are unrealistic. The real level in 1990 for the national government, government corporations and local government units, is just about the level the public sector spent on construction along in 1982, not including durable equipment. In terms of ratio or proportion to GNP, the target for the period 1988-1992 is just about 5.0 percent on the average, below the 7.7 percent which we had reached in some years in the past. In terms of growth rate, the public investment target implies about a 20 percent real growth. In 1976 government construction alone grew by 76 percent. Thus the absorptive capacity of the economy is there. It is just a matter of improving implementation." (Sec.S. Monsod, "RP-IMF Talks: The IMF Prescription is Bad for Us," 4 Dec. 1988, Philippine Daily Inquirer.)

7. In the March 1987 negotiations with commercial creditors, the Philippines obtained interest-rate terms of 7/8 over LIBOR (London Interbank Offer Rate). The comparable spreads for Mexico (October 1986) and Argentina (April 1987)

were 13/16 over LIBOR. In the sequence of negotiating strategies represented by restructuring, lower spreads, new money, and more recently debt-reduction, it has been often noted the Philippine panel's strategy has always been one step behind Mexico's.

8. Though its construction was completed, the Bataan nuclear plan was later revealed to have been built along an earthquake fault and is nonoperational up to the present. Nonetheless, calculated on a daily basis, current interest payments on the loan run up to $355,000.

9. Before his appointment as CB head by Marcos, "Jobo" Fernandez was head of Far East Bank and Trust Co. (FEBTC), a large Philippine commercial bank. In 1981 FEBTC in turn acquired an 87 percent share in Private Development Corporation of the Philippines (PDCP). PDCP was an important channel for World Bank and ADB loans, which were lent to private sector borrowers. The president of PDCP was Jayme, later appointed finance secretary, and its chairman was Roberto Villanueva, subsequently named by Mrs. Aquino as head of the Philippine Assistance Programme (PAP).

10. Curiously enough, it is the economic ministry which seems more sensitive to the political implications of economic measures. In her confidential letter to the president objecting to the March 1989 Letter of Intent, the NEDA head warns of the "political difficulties which the proposed program's implementation may generate." ("Monsod Worried over LOI Impact", Philippine Daily Inquirer, 17 March 1989).

11. Quoted from *Philippine Daily Inquirer*, 17 March 1989.

12. Under the Letter of Intent, the public sector borrowing requirement (roughly the national government deficit plus deficits of government corporations) is to be reduced from an actual 2.4 percent of GNP in 1988 to 1.6 percent in 1990. The assumed growth of exports is 10 percent, compared to an historical 5.7 percent per annum between 1986-88.

13. However, even granting that the unlikely amount of $10 billion over five years is achieved it would not be sufficient to wipe out the net-resource outflow from the debt. There is also a popular perception, albeit officially denied, that donor-enthusiasm for the PAP is linked to the retention of the U.S. military bases in the Philippines.

14. "...Candidates for national offices need votes, which local leaders with their primary hold upon the loyalty of the rural electorate can deliver. Local leaders in turn need money to do favours for their followers, and this the candidates for high offices can supply. Local leaders also need a constant supply of public works projects and services for their localities. Holders of high elective offices such as senators and congressmen, from whence come most such benefits in the Philippines, can affect the supply of projects and services. The result is a functional interdependence of local, provincial, and national leaders..."[Lande 1964:82].

# References

Alburo, F. et al., *Economic Recovery and Long-Run Growth: An Agenda for Reforms*. (Makati: Philippine Institute for Development Studies, 1986).

Bello, W. et al., *Development Debacle: The World Bank in the Philippines*. (San Francisco: Institute for Food and Development Policy, 1982).

Bhagwati, J. "Directly Unproductive, Profit-Seeking (DUP) Activities." *Journal of Political Economy* 90, 1982a, pp 988-1002.

Bhagwati, J. ed. *Import Competition and Response*. (Chicago: University of Chicago Press, 1982b).

Brecher, R. "Comment on Findlay and Wellisz." in Bhagwati (ed.).

Broad, R. *Unequal Alliance, 1979-1986*. (Quezon City: Ateneo de Manila Press, 1988).

Canlas, D. et al., *Towards Recovery and Sustainable Growth*. (University of the Philippines School of Economics, mimeographed, 1986).

Carroll, J. *The Filipino Manufacturing Entrepreneur*. (Ithaca: Cornell University Press, 1964).

Daroy, P., A. de Dios, and L. Kalaw-Tirol. eds. *Dictatorship and Revolution: Roots of People's Power*. (Metro Manila: Conspectus, 1988).

Diokno, M.S. "Unity and Struggle." in Daroy et al., eds.

de Dios, E. ed. *An Analysis of the Philippine Economic Crisis: A Workshop Report.* (Quezon City: University of the Philippines Press, 1984).

_____. "Protection, Concentration, and the Direction of Foreign Investments." *Philippine Review of Economics and Business.* Vol. 23, Nos. 1-2, 1986.

de Dios, E. and M. Montes. "A Perspective of the Philippine Economy." (MS), 1986.

Fabella, R. "Trade and Industry Reforms in the Philippines, 1980-1987: Performance, Process, and the Role of Policy Research." (MS), 1989.

Ferrer, R. "Political Economy of the Aquino Regime: From Liberalism to Bureaucratic Authoritarianism." *Economic and Political Weekly.* July 30, 1988.

Findlay, R. and S. Wellisz. "Endogenous Tariffs, the Political Economy of Trade Restrictions, and Welfare." in Bhagwati, J. ed. *Import Competition and Response.* (Chicago: University of Chicago Press, 1982).

Krueger, A. "The Political Economy of the Rent-Seeking Society." *American Economic Review* 64, 1974, pp. 291-303.

Lande, C. *Leaders, Factions and Parties. The Structure of Philippine Politics.* Monograph Series No. 6, 1974, Yale University Southeast Asia Studies.

Lind, John. *Philippine Debt to Foreign Banks.* (California: Interfaith Committee on Corporate Responsibility, 1984).

Montes, M. *Stabilisation and Adjustment Policies and Programmes. Country Study: The Philippines.* (Helsinki: World Institute for Development Economics Research (WIDER), 1987.)

_____. "The Business Sector and Development Policy." in Ishii et al., *National Development Policies and the Business Sector in the Philippines.* (Tokyo: Institute of Developing Economies, 1988).

Nemenzo, F. "From autocracy to elite democracy." in Daroy et al., eds. pp. 221-268

Soriano, J. "The Return of the Oligarchs. A Preliminary Analysis of the Philippine Congress." *Political Clans and Electoral Politics.* Institute for Popular Democracy. (MS), 1987.

Thompson, M. and G. Slayton. "An Essay on Credit Arrangements Between the IMF and the Republic of the Philippines: 1970-1983." *Philippine Review of Economics and Business.* 22(1-2), 1985, pp. 59-82.

# Japan

## The Administrative Reform, Restructuring and Economic Planning in Japan

*Yoichi Okita*

## 1. Introduction

Japan's most conspicuous political-economic problems of the 1980s are the issues of administrative reform, external imbalances, and tax reform. The last issue, still being examined in the Diet, has not yet reached a stage sufficiently settled to permit overall evaluation, while at the same time being a subject so broad as to defy treatment in a paper of this length. The problem of external imbalance per se is certainly an important one, but such an economic analysis is not the task at hand. What will be examined, in the context of the large current account surplus, is the formulation of policies for "restructuring" the economy. This subject, and the Administrative Reform, are closely related to economic planning, since all of them are essentially maps for future policies. Thus, chronologies for economic planning, the Administrative Reform and the restructuring program (the Maekawa Reports) will constitute the main portion of this paper.

In Section 2, I will present an introduction and an overview of the role of economic planning in Japan in order to lay a foundation for the succeeding sections. Hence the discussion in this section is not limited to the characteristics of plans in the 1980s.

Section 3 is an examination of one plan formulated in 1979. I will explain the revision process in order to demonstrate how complaints concerning overly optimistic forecasting in economic planning originated. The experience of this plan also prefigures the changing nature of later economic planning.

In Section 4, I will discuss Prime Minister Nakasone's unique approach to the economic plan, as signalled by the disappearance of the word "plan" from such titles. Some features of his so-called "Outlook and Guidelines" will also be presented.

Section 5 will provide an overview of the Administrative Reform, focusing on the institutional structures and the procedures, composition and achievements of the steering committee, thus illustrating the strength of business influence.

In Section 6, the background, features and evaluation of the 1986 Maekawa Report plus subsequent work done by the Economic Council will be discussed. It will be demonstrated that this report was a turning point for national focus, and a priority issue in the minds of policy-makers.

In the last section, after comparing the three patterns represented by economic planning, the Administrative Reform and the Maekawa Report, I will present my conclusion that the Japanese government's decision-making system is fundamentally pluralistic.

## 2. Economic Planning in Japan

Several years ago, there was some debate in the United States over the necessity of introducing economic planning. One side argued that government could play an important role in the field of "industrial policy". Compared to this notion of planning, to say nothing of the role of the plan in socialist countries, Japanese plans are relatively weak policy tools for controlling sectors outside of government. Planning in Japan is often characterized as "indicative planning" (as opposed to "imperative planning") and although the view that economic planning is only "decorative"[1] is too extreme, the following description has a certain merit:

Because cabinet approval makes the plan a guiding principle of economic policies during the period, economic planning would confer on the EPA [Economic Planning Agency, the secretariat of the Economic Council], as the principle agency involved, a large grant of authority over decision-making. In reality, it does nothing of the sort.[2]

Before delving further into problems of the Economic Council and the EPA's real function, some basic facts need to be presented. The Economic Council is an advisory board to the Prime Minister which normally convenes at his request to examine a specific policy issue. After a lengthy deliberation, the Council submits its report to the Prime Minister, who then asks the members of the cabinet to approve it. Although, in principle, the cabinet can change or modify the conclusions of the Economic Council, this has never happened. Government officials, after generally heated discussion, settle all points of conflict before the submission of Council's final report.

The Economic Council consists of some twenty to thirty members (if sub-committee members are included, the number often surpasses one hundred) from various fields such as business, academia, labour, consumer groups and semi-public institutions. Although the influence of the Council's steering committee consisting of businessmen and former government officials is important, the pacification of conflicting points is, as described above, primarily accomplished by bureaucrats.

In spite of limitations on its power, the Council provides an important arena for selecting and mollifying conflicting views from varying social groups. Ideally, it would be the place to create a national consensus on many important subjects. The most important issues are, however, quite often politically the most difficult. Although it is easy to point out the lack of bold reform proposals in the conclusions of the Economic Council, no other institution, council, or committee could do better, because they too would labour under the same difficulties that the Economic Council faces.

The Economic Council is invested with another important function. It provides medium-term projections for various indicators including the GNP, the demand components of the gross national expenditure, price indices, sectoral decomposition of output and labour market indicators. These figures are used, or at least referred to, in many of the medium-term plans promulgated by various ministries and agencies.[3] Many of the ties between the national plan and other plans are not binding and are sometimes invisible, but, because it is a convenient source for an authorized GNP growth rate prediction, it has become the common source for many branches of central and local government.

From a political viewpoint, the most important figures are the target for cumulative public investment during the plan period and the total government investment allocated to some fifteen public work programs aimed at ports, airports, roads, railways, sewerage systems, flood control, public sector housing, and so on. (This aspect of economic planning has been frozen under the last two plans, a point to be discussed in detail below.)

Trezise's comment provides a convenient point of departure for discussing the relative merits of economic planning. He writes:

> For one thing, the plans have no binding force on anyone. Private investors may at times have responded to the announcement effect of the plans, but hardly according to the envisioned targets. After The Plan to Double National Income was unveiled in December 1960, private plant and equipment investment in the next year alone almost reached the level projected for 1970.... As for the public sector, the volume of government capital formation is decided in the annual budgetary process, which is influenced only remotely, if at all, by the plan.[4]

Concerning Trezise's first point, it should be noted that these plans are not aimed at the private sector, because the Japanese economy is not a "planned economy". At the same time, it must be remembered that even in countries where no institutionalized economic plan exists governments try to influence the private sector by demonstrating a "credible" attitude toward certain economic policies, or by manipulating its expectations. This also indicates that the announcement effect is a matter of concern for authorities in many other countries. Since Japan's Prime Minister has fewer occasions to deliver speeches to achieve this effect than, say, the President of the United States, economic planning can be a rather important tool for communication and persuasion.

Turning to Trezise's second point, given the fact that the Ministry of Finance was quite eager to freeze the process of government investment allotments under the last two economic plans, we can easily infer that this has an important influence on the budgetary decision-making process. In addition, economic plans have some influence over the annual forecast, or more literally, the government's annual outlook, as prepared by the EPA. John C. Campbell reports that the Ministry of Finance (MOF) has its own forecast[5] and, based upon this observation, Trezise concludes that once a compromise between the EPA and the MOF is obtained on the target growth rate figure, "the pertinence of the medium or long-term estimate becomes questionable."[6] It would be more appropriate to describe such a compromise as the coordination of various views held by all the ministries and agencies, including the EPA and the MOF. Since the EPA is designed to be a coordinator, it is only natural that such compromises take place. It is not the EPA that is yielding, but all other ministries, including the MOF and MITI (Ministry of International Trade and Industry), since, as a mediator, the EPA has nothing to yield.

Of course, there is always the problem of the quality of coordination. "Add two and divide by two" style compromises may not be the best approach. The problem of how to achieve desirable compromises is inherent in any kind of political decision-making process. It also depends on the leadership of the Prime Minister. Fortunately, political and administrative systems and economic performance in Japan (and in many other countries) are normally resistant against inferior coordination efforts (the second best solution).

Under abnormal conditions, strong leadership may be needed. In Sections 4 and 5 below, the situation under strong leadership will be examined, but before that, some recent changes in the role of economic plans will be discussed.

## 3. Economic Planning and the Second Oil Crisis

On August 10th, 1979, the Ohira Administration approved the "New Seven Year Economic and Social Plan" (hereafter referred to as the Seven Year Plan), proposed by the Economic Council. The date of approval is significant, as it was during the turmoil of the second oil crisis. In retrospect, it is easy to point out that the adjustment process caused by changing energy and material costs should have been anticipated. One might say that in light of the experience of the first oil crisis, the Economic Council and the EPA should have taken a very cautious attitude when plotting forecasts using the established method.

The EPA did try to change the econometric model for economic planning some years after the first oil crisis, so that some kind of supply-side economics and the effect of changes in energy price could be incorporated. Although the resulting model was very sophisticated by any standards, it did not employ techniques used by the anti-Keynesian school, such as the rational expectation theory. In any case, anti-Keynesians oppose economic planning and forecasting with econometric models.

When the Seven Year Plan was being prepared, the EPA and the Economic Council could not foretell how much higher crude oil prices would rise, nor did they know whether what they were observing constituted a crisis similar to that experienced in 1973-74. It was, therefore, difficult for them to determine an appropriate timing for the initial year of the projection, let alone adequate assumptions concerning external factors.

Still the EPA and the Economic Council did learn from the first oil crisis. Forecasting errors in the plan adopted in 1979 were much smaller than those of plans prior to 1974. (See Table 1).

In spite of lowered growth rate figures, the Seven Year Plan was criticized for projecting growth rates that were too high. In January 1980, the plan was in a sense revised in the so-called "follow-up procedure" report, essentially an annual review, when the Economic Council announced a new prediction of a 5.5 percent growth rate. In this report, there was another important revision. In the original plan, the introduction of a "general consumption tax" (a type of value-added tax) was proposed, but due to unfavourable election results in December 1979, the idea was officially abandoned in the follow-up report.[7]

In the next follow-up report of January 1981, the growth rate projection was left unchanged, but a modification that seriously altered the conventional function of economic planning was introduced. This was the revision of the target figure for total cumulative government investment. Originally, the Seven Year

Plan had a target of 240 trillion yen (roughly $1.8 trillion) cumulative expenditures on public sector investment programs for that period. Concern over the growing budget deficit, however, made the argument for fiscal consolidation more persuasive, leading to the revision of the target down to 190 trillion yen. Although it was not clear whether it was legitimate to change such an important item without introducing a new plan, there was no criticism of this revision.

Table 1
Real GNP Growth Rates: Forecast and Actual Rates

| Title (Administration) | | Forecast | Actual |
|---|---|---|---|
| New Economic and Social Development Plan (Sato) | 1970-75 | 10.0% | 4.9% |
| Basic Economic and Social Plan (Tanaka) | 1973-77 | 9.4 | 3.5 |
| Economic Plan for the Last Half of the 1970s (Miki) | 1976-80 | 6.0 | 4.0 |
| New Economic and Social Seven Year Plan (Ohira) | 1979-85 | 5.7± | 4.1 |
| Outlook and Guidelines (Nakasone) | 1983-90 | 4.0± | 3.9 (83-86) |

In January 1982, the third follow-up report again lowered the projected growth rate, this time to 5.1 percent. This, plus the frequent revisions of other important aspects, indicates that the concept of the plan was getting much more flexible than in the era of rapid economic growth.

In spite of this, some leaders of the business community, as well as others on the Special Committee for the Examination of Administrative Systems (hereafter referred to as the SCEAS), established under the Administrative Reform, were said to have complained about the rigidity and excessive optimism in the forecasted growth rates which, according to them, provided a basis for unchecked expansion in government expenditures.

In the final (fifth) report of the SCEAS, published in March 1983, they advocated flexible revisions of government plans, proposing the following:

> The figures projected in government plans should not be regarded as fixed; rather, they should be revised in a flexible manner so that the plans can be adjusted adequately to meet the changes in surrounding conditions. In the periodic reviews, efforts should be made not only to adjust the projections of economic indicators, but also to revise policy targets when necessary. Furthermore, emphasis must be placed on checking the degree of concrete progress made towards realizing the proposals presented in the plan.

In a sense, The Seven Year Plan follow-up procedure preempted the SCEAS proposal. Still, the above passage indicates that the situation was not yet fully satisfactory. The influence of the SCEAS on economic planning will be examined again in Sections 4 and 5.

After experiencing repeated misfortunes with the Seven Year Plan, the EPA and the Economic Council, in the summer of 1982, resolved to introduce a new economic plan to avoid the continual criticism of the previously accepted procedures. In addition, the stabilization of oil prices, albeit at a quite high level, may have reassured planners that it was relatively safe to make economic predictions, since exogenous or environmental variables were more predictable than they had been in the midst of the second oil crisis.

In July 1982, Prime Minister Suzuki requested a 5 year plan from the Economic Council. Two significant factors indirectly affected the course of Council deliberations. First, the activities of the SCEAS were gathering momentum just as the media began to scrutinize this new advisory system. Second, it was becoming clear that the goal of "fiscal consolidation without tax increase," the emergent theme during the early stages of SCEAS discussion and its indispensable principle, would be difficult to achieve by the target year of 1984, due to the reduced increase of government revenue following the second oil crisis. Setting new goals for fiscal consolidation would entail controversy over consistency between this and projections in the new economic plan, thereby complicating efforts to justify both the targets and the economic plan simultaneously. Although in principle the Economic Council, with the leadership of the Prime Minister, was positioned to lead budget authorities toward rectifying the two targets, the Ministry of Finance seemed to have had a certain influence.

## 4. Economic Planning under the Nakasone Administration

In November 1982, just as deliberations for a new economic plan had commenced, Mr. Nakasone won the position of Prime Minister. The EPA, as secretariat of the Economic Council, hoped to complete the plan by late 1982, following the pace set by previous plans. That expectation was not met, however, due to the new administration's desire to promulgate its own unique style of economic planning.

In January 1983, Mr. Nakasone asked the Economic Council first to extend the term of the plan beyond five years, and second, to make its projections and guidelines more flexible so that rapidly changing conditions could be accommodated. The request was expressed thus:

> In the coming years, the economy and society of our country will be affected by many uncertain factors, which make it difficult to project or target future courses in a rigid manner. In view of the need for an economic plan responsive to these kinds of changes, I would like to request that the members of the Economic Council present an outlook on our economy and society and a guideline for the economic policy management under a span longer than a five year period.[8]

The Economic Council responded by extending the period of the plan to 8 years and by seeking a plan format that could accommodate more flexible projections.

In August 1983, the Council finalized the new economic plan, titled "Outlook and Guidelines for the Economy and Society in the 1980s", which they submitted to the Prime Minister. It was adopted by the Cabinet on August 12. This plan, according to a summary by an EPA official engaged in its preparation, had characteristics different from all previous plans.[9]

First, the plan was given more flexibility, so that it could cope with uncertain and changing conditions. It was deliberately given the name "Outlook and Guidelines" instead of "Plan" to express this feature.

Second, the following four points were significantly placed at the top of the list of recommended policies to indicate that they were the priority features: (i) administrative and fiscal reform (the most urgent task); (ii) new forms of economic growth supported by further sophistication of the economic structure, such as the development of high-tech sectors and information-related sectors; (iii) an emphasis on the role of private sector vitality; and (iv) greater international cooperation through such means as increasing development aid and contributions to international economic systems and better utilization of the vitality of the private sector.[10]

Third, the following five themes were emphasized as general guidelines for deciding concrete programs: (i) the promotion of peaceful and stable international relationships; (ii) the development of an economy and society full of vitality; (iii) the maintenance of a safe and comfortable life style; (iv) the achievement of moderate economic growth consistent with full employment, price stability and equilibrated external balance; and (v) the reform of the administrative system and fiscal consolidation.

Fourth, to cope flexibly with possible environmental changes, a system of annual re-examination named the "revolving plan", was introduced. The term "rolling plan" was not used because the final year of the plan was fixed at fiscal 1990.

In contrast to the features outlined by the government official, the following characteristics appeared rather starkly stated in my own review of the plan:

1. An emphasis on the necessity to utilize free market mechanisms as much as possible;
2. Fewer types of projections and guidelines (nominal and real growth rates, CPI and WPI inflation rates, and the unemployment rate were the only figures presented);

3. An emphasis on the uncertainty and difficulty of prediction, as indicated by the stress of annual review (revolving), a stronger form of annual follow-up than used for previous plans;
4. The abandonment of the targeted total value of public investment during the period of the plan; and
5. Repeated mention of the need for administrative reform in response to then on-going activities of the SCEAS.

The portion of "Outlook and Guidelines" that embodies these concepts is the very beginning of the text, a sub-section titled "Fundamental Role and Philosophy". It reads as follows:

> Japanese society is based upon the principle of free competition and market economy. An economic plan compiled under this premise is not designed to provide detailed regulations governing all economic and social fields. Nor is it meant to be implemented rigidly and forcibly.
>
> Rather, the economic plan is fundamentally aimed at: (1) clarifying prospects for a desirable and realizable state of the economy and society; (2) defining the basic economic policies to be pursued by the government in the medium and long-term periods, and spelling out priority policy goals and ways to attain them; and (3) providing guidelines for household and business activities.[11]

It should be noted that the philosophy outlined above, and attendant policy features, were fashioned into the target of fiscal consolidation and administrative reform. Naturally other goals were not emphasized quite as much. Kiichi Saeki, a member of the Economic Council, recalls the process of deliberation with a somewhat despondent tone:

> Quite a number of people seemed to be dissatisfied with the recent "Outlook and Guidelines for the Economy and Society in the 1980s". Some newspapers criticized it severely. I myself have many complaints. Bluntly stated, there is no plan, just meaningless words. It is not clear enough nor persuasive enough to deserve the title of "outlook" or "guidelines". Everything, from the targets to policy tools, are obscure. . . .

> I do not understand why we had to be so sensitive to using the word "plan". It was only October 1982, when a special issue of the *ESP*, to which I made a contribution, was devoted to the theme of the economic plan. Now the word "plan" is almost taboo. Is this "Outlook and Guidelines" really an economic plan? . . .
>
> Being a member of the Sub-Committee on International Economy, I am most anxious over "achieving an internationally harmonious external balance through growth centred on domestic demand." Although this theme is emphasized as one of the five major policy targets in the section on Japan's contribution to the development of international economy and society, it lacks substance and clarity. The Sub-Committee on International Economy has stressed the need for domestic, demand-led growth, but has not attempted to delineate the means to achieve it because we thought that was mandated to other sub-committees. In the end, however, the "Outlook and Guidelines" does not indicate how to achieve domestic, demand-led growth without jeopardizing administrative and fiscal reforms, the number one priority in it.[12]

Although it is debatable if the whole exercise of the "Outlook and Guideline" was meaningless, Mr. Saeki's point on the necessity for structural shift from external demand to domestic demand was a noteworthy insight. His statement prefigures the Maekawa Report of three years later. During the intervening three years, the Japanese government and perhaps even the public at large were able to focus on administrative and fiscal reforms without much disruption by external problems.

## 5. The Administrative Reforms— Mr. Doko and Mr. Nakasone

The movement for administrative reform was led primarily by the activities of the Special Committee for the Examination of Administrative Systems (SCEAS) and the Special Council for the Implementation of Administrative Reforms.

The SCEAS was given the same name as a committee established in 1961 and terminated in 1964. This committee made some unique proposals, such as introducing modernized and rationalized management into administrative procedures. Although the committee established in 1980 has been often called "the Second Special Committee", it can be regarded as an independent phenomena responding to the central government's increasing budget deficits and the prospect of tax increases. One official in the secretariat of the Special Committee described the difference between the second and the first committee as one of system reform versus technical reform.[13] Put more precisely, the second committee concerned itself with the more fundamental issue of determining the most desirable content, boundaries and size of the public sector and government controls, while the first one was concerned with the methods and procedures of the government's daily business. Owing to the second committee's approach, such radical reforms as the privatization of the Japanese National Railway and the Telephone and Telegraphy Authority, as well as some deregulation, was possible.

The SCEAS consisted of nine official members. Its chairman was the late Mr. Doko, one of the most influential leaders of the business community and a former president of the Keidanren (Federation of Economic Organizations), famous for his simple lifestyle and his ability to resuscitate companies in crisis. Two businessmen, two labour union leaders, two former government officials, one journalist, and one academic made up the rest of the committee. In sum, this composition favoured, not ideally but sufficiently, the business community.

There were others who participated in the process. Six Councillors acted as supplementary members of the committee, but their participation was rather infrequent. The title of "expert" was given to twenty-one persons engaged during the discussion of specialized fields. Fifty-five consultants were employed at the subcommittee and expert meetings and were accorded generally the same treatment as the experts. Although some argue that the presence of so many former government officials in the subcommittee necessarily weighted the committee to the interest of those sections of the government connected to various pressure groups, the outcome, discussed later, did not bear out this position.

The bill establishing the SCEAS was passed in late 1980. The Committee held its first meeting on March 16th, 1981 following the government's selection of its members and the Diet approval of the appointments. At this meeting, Prime Minister

Zenko Suzuki asked the Committee to provide basic reform proposals for administrative systems and for the modus operandi of central and local governments, together with a study of detailed measures to achieve reasonable operational and administrative structure. He stressed that the committee was the authoritative organ for the study and examination of administrative matters and for designing their reforms. He also requested that the committee make some concrete proposals aimed at smoothly settling the budget deficit problem, utilizing such measures as cutting expenditures and simplifying government and streamlining administrative offices. He explained that such efforts were essential to the return to a healthy budget and administrative systems.

At the same meeting, Mr. Nakasone, then Minister for the Administrative Management Agency, in his opening statement to the Committee, delineated its basic goal as two-fold. One goal would be the realization of "simple and efficient administrative systems", appropriate to a period of moderate growth, and based on thoughtful review of the role of central and local government. The other would be the design of a better institutional framework for the public sector, accomplished by programs for its realization. Since the law establishing the SCEAS was rather vague and overly general, Mr. Suzuki's and Mr. Nakasone's statements may be regarded as practical mandates.

One can observe from these statements that the Committee was assigned a task that overlapped with the Economic Council and economic planning. The reform of administrative systems, which quite often has important economic effects, is in fact a part of economic policy and the design of future structure is also a function of economic planning.

One month after its inception, the Committee determined its program, publishing a document called, "Basic Subjects for Full-Term Study and Urgent Problems for Immediate Consideration". Although this document did not strictly guide discussion, it initiated work towards the submission of the Committee's first report in July. In this report, as the Prime Minister requested, the Committee proposed a number of stringent measures for cutting expenditures, such as limiting the budget of every ministry to the same level as the preceding year (the so-called "zero percent ceiling rule"), an across-the-board cut in subsidies and preferential tax treatments, some detailed proposals for reining in unleashed increases in social and medical care programs, increased contributions to social pension schemes and a ceiling on government hiring. Some of these proposals were reflected in the 1982 budget. Perhaps the greatest impact of the first report was on public opinion; it became the focus of keen attention from journalists and the general public.

Another important result of this discussion was the establishment of four sub-committees, whose mandates set the course for subsequent discussion in the main committee (See Table 2). Their detailed and fairly concrete agenda served to stimulate public curiosity. Furthermore, the May 1982 reports from these four sub-committees served to attract journalistic attention to the Administrative Reform.

The First Sub-Committee proposed, as an immediately feasible solution to the problem of adequate government services and tax burdens, the principle of "fiscal consolidation without tax increase". The *medium-term* goal was to contain the growth of central government budget expenditures at a level below the nominal GNP growth rate, and to keep the ratio of general government expenditure to GNP at around the current level of 35 percent. The *long-term* goal targeted the ratio of total tax and social security burdens to national income at a level that would not rise beyond European averages of 50 percent.

The most interesting suggestion of the report of the Second Sub-Committee was that the power to coordinate branches of government should be strengthened to combat sectionalism between ministries. It was ironic, that while being critical of some functions of the Economic Planning Agency (such as its long-term projections of macro-economic indicators and their targeting of government investments), they were eager to strengthen the coordinating aspect of the agency. This idea, however, did not end up affecting the EPA. Instead, it was later modified, and became the impetus behind the establishment of the General Administration Agency.

The Third Sub-Committee issued a report on its study of local government. It suggested reforms reassigning functions of the central government to local government, as well as reforms aimed at specifics within local government itself.

## Table 2
## Questions Before the Sub-Committees

The First Sub-Committee

i) What is the philosophy that necessitates administrative reform? What are the are the medium and long-term visions of administration?

ii) What is the real task of the public sector and how should critical problems, such as agricultural policies, social security, housing and land policies, energy policies, science and technology policies, overall national security, tax systems, and so on, be handled?

The Second Sub-Committee

i) Review the organization of all ministries and agencies.

ii) Examine how to reinforce synthesizing and coordinating functions within government.

iii) Review the process of budget formation, appropriation and off-budget programs (the Fiscal Investment and Loan Program).

iv) Improve government ethics; create new rules for government employment.

v) Examine how to control and open administrative information.

The Third Sub-Committee

i) Determine the proper division of functions and resource distribution between central and local governments.

ii) Review protective policies and regulations.

The Fourth Sub-Committee

i) Establish rules demarcating the public and private sector, study the possibility of privatizing or rationalizing the three public corporations, the five government undertakings, and other direct government operations.*

ii) Review institutional forms of semi-governmental corporations (government supervised corporations or "chartered corporations").

* Note: The three public corporations are: the Telephone and Telegraph Public Corporation, the Japan National Railway, and the Japan Tobacco and Salt Monopoly. The five undertakings are: the Postal Operation, the National Forestry Operation, the Government Printing Office, the Mint, the Industrial Alcohol Monopoly. Some examples of semi-governmental corporations include: the Japan Development Bank, the Japan Ex-Im Bank, and the Japan Housing Construction Corporation.

The Fourth Sub-Committee proposed reforms for three public corporations. This was perhaps the most significant outcome of all the sub-committees' proposals. The basic strategies it recommended were to give these corporations the freedom of self-determination and to push management and employees toward significantly improving efficiency. According to the report, one of the requisite conditions for improving serious deficits in the Japan National Railway were privatization and the dismembering of the system into regional companies.[14] They also reported that rapid changes in information technology would make the Telephone and Telegraph Authority obsolete unless private management, capable of rapid adjustment, was introduced.

Prior to the publication of these reports in February 1982, the main committee composed a second report, concerning deregulation. In July, two months after the sub-committees' reports, the main committee's third report, called the "Basic Report", was released. This report was a compilation of the sub-committees' reports, and did not contain any major new ideas.

In November, an important political development occurred, greatly affecting the course of the Administrative Reform. Mr. Nakasone, the very person in charge of the reform, was elected the Prime Minister after the resignation of Mr. Suzuki. It would seem he had found that the Administrative Reform had good political potential. The Reform thus became an important feature of his administration. Mr. Nakasone promptly initiated legislation for Japan National Railway Reform Act, which led to its privatization.

In the final stage of SCEAS deliberations, few new recommendations on essential matters were added. Rather, efforts were concentrated on substantiating the abstract concepts discussed in the Basic Report. Still, a number of points that emerged during this stage need to be mentioned. The first was a set of additional reform proposals concerned with streamlining the postal service, the national forestry operation, national hospitals, and numerous chartered corporations. The second was the reorganization of offices in the six ministries and two agencies of the central government. The third proposed curtailing the local dispatch of central government ministries.

Other proposals that surfaced during the final stages of the deliberations dealt mostly with strengthening measures to cut expenditures, such as more stringent principles for reducing the number of government officials and concrete programs for cutting various subsidies.

One important proposal not developed into a final recommendation, but discussed intensively, was the treatment of the postal savings system. Private banks were naturally anxious to prevent the expansion of this system, which tended to put them at a tax disadvantage.[15] Some opponents argued that private banks were also protected by the de facto cartel created by regulations and administrative guidelines.[16] The compromise settled upon was to combine supportive language for continuation of the postal savings system in general, while at the same time placing limits on one particular type.

The Fourth and Fifth Reports published in February and March of 1983, constitute the committee's final recommendations to the Prime Minister. On March 15th, the committee was dissolved, as scheduled by law.

The government had authorized many of the Committee's proposals at an earlier stage by preparing the necessary legislation. In May 1983, the law establishing the new Council for the Implementation of Special Administrative Reform was passed by the Diet. It was to discuss all issues addressed by the SCEAS, monitor the implementation process for the reforms proposed by the SCEAS, and add concrete and detailed proposals. In September of that year a special session of the Diet was held to discuss the bills effectuating administrative reforms. Twenty-seven bills were passed by the Diet during this so-called Administrative Reform Session and the next session, thus making possible the realization of the proposals.

The Council continued to be active with significant further results in areas such as local government reforms. Its achievements were important enough to describe its work as successful. The reforms and privatization of three public corporations, the reform of the social security systems, the merging of some government offices into the General Administration Agency, and the relaxation of some 35 regulations are particularly noteworthy. One might argue that the Administrative Reform was partially a failure, because the most critical target of fiscal consolidation, defined as eliminating the issue of government bonds not matched by capital accumulation (called Current Deficit Financing Bond or Deficit Financing Bond) was not achieved. It may be, however, that defeat here was more that compensated for by gains on other fronts. In any case, the expenditure side of the budget during the first half of the 1980s was rather strictly controlled, with expansion much more limited than before.

Why did the SCEAS system work so well and to whom does the credit belong?

First, the tactics taken by Chairman Doko and a few of the business leaders were quite effective. Mr. Doko presented four appeals to Prime Minister Suzuki at the outset of SCEAS's activity. With the Prime Minister's consent, these four points became the agreement binding him and his government:

1. The Prime Minister pledges to bring to fruition committee proposals, availing himself to the full authority of his office and over his party;

2. The Prime Minister shall champion the ideas behind SCEAS, to realize "smaller government" and "fiscal consolidation without tax increase";

3. The Administrative Reform shall include not only reforms for the central government but also for local governments and other governmental bodies; and

4. The Prime Minister shall encourage vitality in the private sector by eliminating deficits in three crucial areas (the National Railway, the Rice Control Program, and the Social Health Insurance Program), by cutting the number of semi-governmental corporations and through privatization.

Prime Minister Suzuki, too, was desirous of some strong focus for impeding the growth of the fiscal deficit. Mr. Nakasone, Mr. Suzuki's successor, was even more determined to make the Reform his principle campaign issue, since he had initiated the effort as head of the responsible agency. Even without the accord with Mr. Doko, these two would have wielded leadership effectively.

The business community rallied behind the Administrative Reform, perhaps catalyzed by the past failure to prevent corporate tax increase in fiscal year 1981. It succeeded in occupying key positions in the SCEAS, and considerably influenced the course of its discussion. Even outside of the committee, it worked to persuade politicians and the public.

The SCEAS also succeeded because it attracted the attention of the media by deliberately leaking information from closed-door discussions. In addition, the frequent publication of reports, owing to the multiplicity of stages and sub-committees helped to focus public interest. The theme of "punishing evil government officials who steal tax money" has always enjoyed popularity with the media. Harkening to this time-honoured theme, many newspapers gave extensive coverage to the need for Administrative Reform.[17]

The committee itself was given the freedom to choose its own style and fields of deliberation. This freedom allowed it to focus on the problems it felt were of the greatest concern and to avoid being bogged down with external considerations. The committee was lucky in that it began its work before problems such as the trade imbalance had grown to the point of creating international friction. It was also fortunate that international economic thought, as represented by Reaganomics and Thatcherism, had begun to shift away from Keynesian economics. Quite often, the U.S. government endorsed and at time praised Japan's "small government" policy even though it may have weakened domestic demand to some degree.

Finally, committee conclusions did not need to be submitted to the Cabinet for approval. This is clearly different from the procedure for economic planning. If they had been subject to examination by the Cabinet, every single branch of government would have had, in effect, a right to veto them.

This is exactly the problem that has plagued the Economic Planning Agency. As an office engaging in arbitration, the EPA is forced to balance the arguments of one ministry against another, a difficult operation when those who are being coordinated have the right to veto the resulting compromises at the Cabinet level. The SCEAS was free of this problem, but they did have to concern themselves with mobilizing politicians in order to develop their ideas into enforceable laws and directives.

## 6. The Maekawa Report and Moves toward Restructuring

On April 7, 1986, the group mandated by Prime Minister Nakasone and chaired by Mr. Maekawa, former Governor of the Central Bank, submitted its report. This committee was established to study economic policies in light of the rapid increase in the 1984 and 1985 trade surplus. In 1986, the current account surplus reached 3.6 percent of the GNP. The report explained the need for policy change through "restructuring" as follows:

> It is imperative that we recognize that continued large current account imbalances will create a critical situation not only for the management of the Japanese economy but also for the harmonious development of the world economy.
>
> The time has thus come for Japan to make a historical transformation of its traditional policies on economic management and in national life-style. There can be no further development for Japan without this transformation.

The change of atmosphere surrounding economic policy became apparent by the time the Special Council for the Implementation of the Administrative Reform was terminated. In June 1986, just two months after the Maekawa Report, the Council published its final report and was dissolved. Although a new council under the same name was established in February 1987 for follow-up activities, the initial dissolving of the committee in 1986 effectively marks the end of an era. An editorial in *Nikkei* (The Japan Economic Journal) commented that the administrative reform was now faced with an "adverse wind", namely conditions not foreseen when the effort had been put into motion. The most important factor was the external imbalance. The author of the editorial was quite sympathetic to the idea of small and efficient government and recognized that there was much left to be done. Yet, even this editorial admitted that the domestic balance of savings and investments, exacerbated by insufficient government consumption and investment, was a contributor to the imbalance and that it was time to design concrete measures for coping with changes in external conditions.

Another indication that times had changed was the tone of the Council's final report. In one section, it was admitted, albeit very abstractly, that, when there was sudden change in external conditions, the government could relax its principle of austerity and that the effective tax rate could be allowed to rise if it would be effective in diminishing some unfairness in the system.

The composition of the Maekawa group also embodied certain changes. Out of the seventeen persons in it, three were former officials of the Economic Planning Agency. This was quite a contrast to the representation on the SCEAS, which had only one sub-committee member who had been a member of the EPA.

Turning to the contents of the report, it is important to investigate why the term "structure" was emphasized so much. Restating the question, we must see if economic structure was a real factor in the rapid expansion of the surplus.

The government's 1986 "Annual Economic White Paper" contained an analysis of factors contributing to the increase in the trade surplus during 1982 to 1985. It was estimated that 40 percent of the increase was the result of the gap between domestic demand growth rates and foreign demand. The remaining 60 percent was ascribed evenly to the influence of exchange rates and differences in import and export elasticity between Japan and the rest of the world. Here, the last factor may be called a "structural" factor, because such elasticity is a coefficient in import and export equations. (Econometricians sometimes refer to the set of coefficients as the structure of a model).

In any case, structural factors do not explain the heart of the problem. Leaving out the exchange rate factor, which cannot be easily controlled by Japan, policies affecting the total volume of domestic demand must be the focus if the surplus country is seen as responsible for the surplus. When the Maekawa Report was being prepared, the budget deficit had not yet shown any signs of significant improvement and it was difficult to convince politicians and the Ministry of Finance that some fiscal policy must be employed. Monetary policy was further constrained by high U.S. interest rates. Thus the Maekawa Group had no alternative but to resort to manipulating structural factors.[18]

The report did recognize the importance of increasing domestic demand, however, as indicated by passages such as "[p]romoting the transformation of export-led economic growth to that driven by domestic demand requires that the government put firmly into place domestic-demand expansion policies that have large multiplier effects and will lead to increased private consumption." The concrete measures recommended by the report included neither tax cuts nor public expenditure policies. Aside from a proposal to expand tax deductions for housing purchases and a proposal that can be interpreted as an interest subsidy on housing loans, measures listed in the report can generally be classified as deregulatory. These included easing regulations on housing construction, reductions in total working hours, and improvements in market access while encouraging imports through the deregulation of commerce and distribution. The report showed that the Maekawa group relied on market principles or mechanism.

When the Maekawa Report was published, a meeting of economic ministers announced that they would seek concrete measures to achieve the Report's recommendations by introducing some new meetings, such as the Cabinet and Ruling Party Meeting for the Promotion of Economic Restructuring. In September that year, the Special Sub-Committee for Economic Restructuring was established by the Economic Council. Mr. Maekawa chaired this meeting as well. These actions resemble those taken after the termination of SCEAS in 1982, giving the appearance that significant steps were being taken. There was in fact, a real need for a re-examination of problems by the new Maekawa group. Due to the very rapid appreciation of yen, some people began to fear that restructuring was proceeding too rapidly to have the desired effect. The task of this second Maekawa group was to "redefine the fundamental concept of economic restructuring."[19] Thus, the new Maekawa Report attempted to demonstrate that the yen's appreciation was itself a favourable condition for bettering living standards. "Looking inward, it is questionable whether or not Japan's economic growth is reflected in the quality of Japanese life; housing standards are low, the cost of living high, and working hours long... The people have thus began to wonder whether the yen's strength is reflected in their own standard of living."[20] The report blamed Japan's economic structure for preventing the devalued dollar from being reflected in real standards. The word structure was used to indicate private and social systems, regulations and other *fixed* things.

As to specific recommendations, few items were entirely new, but, if one reads between the carefully worded lines, one can see that their expression was strengthened.

> Fiscal and monetary policy has an important role to play in seeking to achieve economic growth led by domestic

demand, and it is especially important to make use of fiscal policy's resource redistribution functions.

While observing the basic spirit of administrative and fiscal reform and taking advantage of the progress made thus far, extraordinary and urgent fiscal measures should be taken to stimulate domestic demand, given the current economic situation. Efforts will continue to be made to establish appropriate and timely fiscal and monetary policy management.[21]

In May, just after the completion of New Maekawa Report, the government adopted a comprehensive package to stimulate domestic demand, called "The Emergency Economic Measures". This included a special disbursement of five trillion yen for public investment, income tax cuts of about 1.5 trillion yen and an expansion of public housing loans. The total extent of these measures reached almost two percent of the GNP. In a sense, the ideas discussed in the Maekawa Report and the Economic Council's report were materialized in this package. There were many other factors besides the recommendations in these two reports that influenced the decision to take such drastic measures. One clearly apparent factor were demands made by other countries, which garnered particularly keen attention from policy-makers in the pre-Summit season.

Although the platform they popularized relied on a loose definition of structure, the two reports achieved several important gains in economic policies in addition to those described above. They included the reduction of the tax preference on savings, lowering the legal limit on working hours, cutting back domestic coal production, and establishing offshore financial markets.

Many, including foreign observers, evaluated the set of actions taken between the first Maekawa Report and the May 1987 emergency measures as successful. The Japanese economy in 1987 and 1988 has shown remarkable strength in domestic demand. But the reasons for this success seem to be partially attributable to traditional factors, such as monetary relaxation and expansionary fiscal policy measures that cannot be called structural. Likewise, recent increases in imported manufactured goods are not only a result of restructuring and de-emphasizing comparatively disadvantaged sectors, but also a result of the yen's appreciation.

Foreign observers seem to have had excessive expectations about how the government could respond to an announcement like the Maekawa Report. This has led to the view that recent improvements in the Japanese economy are mostly owing to actions based on the restructuring reports. Although the attitude of the Japanese government and aspects of government documents including the Maekawa Reports no doubt effected foreign expectations of the view that restructuring alone could be an effective tool is a fallacy similar to the concepts of "Japan Inc.", "MITI control" or "The Japan Problem". If Japan's economic system were to be conceived as mostly controlled by extra-economical factors, the appeal for structural transformation as *the* national goal would be very promising.

If one regards the situation objectively, it is clear that the proposals embodied in the two Maekawa reports were rather modest, and, given this modesty, the resulting policies and economic conditions were rather impressive achievements. Even though the effort did not have such a fundamental economic effect as to be "structural," it did bring about a shift in national goals and policy objectives from the Administrative Reform to "internationally harmonious" economic performance.[22] One might say that it was not the structure of the economy or industry, but the structure of policy that was altered by the efforts of two Maekawa groups.

In 1988, Prime Minister Takeshita, Mr. Nakasone's successor, introduced a new economic plan which replaced the one adopted five years before. The interval between this plan and the previous one was exceptionally long, given the history of economic planning. On the one hand, this reflected the fact that actual developments were close to the projections in the previous plan, but on the other hand, it may be that Mr. Nakasone's administration's focus on Administrative Reform during this period deflected its attentions from the planning process. The contents of the plan and the chronology of the deliberation process will not be presented here, but one thing needs to be mentioned. The fact that the word "plan" has been restored to the title, albeit modestly,[23] may symbolize the need for an institutional framework different from the SCEAS. Consideration of international problems was comparatively weak during discussions of the domestic problem of Administrative Reform, while an institution like the Economic

Council could transcend domestic interests if the leadership of the Prime Minister is focused on external issues.

## 7. Conclusion

One tentative conclusion from these observations concerning the Japanese advisory system is that the most desirable process for achieving a consensus may vary according to the nature of the problem, essentially whether it is an international problem or a domestic one. Some care must be taken when ascribing too much influence to interest groups, particularly business enterprises, in government policy-making process.

One view on this process is "pluralism," which states that many interest groups are in constant competition, with no one group having any particular advantage over the others. The degree of influence of any one group, say a group of enterprises, on the formation of economic policy depends on public opinion, economic conditions, and the relative vigour of other interest groups. This pluralist view has been challenged by scholars such as Robert A. Dahl and Charles E. Lindblom, who argue that since enterprises are not democratic organizations and since they have the power to influence the national decision-making process, pluralism is always threatened by their pressure.[24] Though space limits proper discussion of this argument, it is enough to say here that the process of the Administrative Reform and economic planning suggest that the Japanese system can be better characterized as pluralistic. Still, aspects of Administrative Reform were coloured by influence from interest groups, along the Dahl-Lindblom lines.

Professor Masahiko Aoki presents a difference dichotomy. One hypothesis is that the state or the government acts as an "interest-mediator," achieving compromise between various (i.e., pluralistic) interest groups. Then as an opposing hypothesis, he presents the view that the government act as a rational decision-maker (paternalistic role).[25] Although he concludes that these two are not mutually exclusive, but rather two facets of the character of government, the view that there are two different types of government is adopted here, since this bi-polar characterization is a convenient tool for handling the cases discussed in this paper.

Although the Administrative Reform does not fit into this schema, the Maekawa report may be interpreted as paternalistic. It must be hastily added, however, that its essential character was not very different from the model of pluralistic decision-making.

Muramatsu, Inoguchi and others describe the Japanese political decision-making process as "quasi-social bargaining nested within the bureaucracy and party politics" which partakes of the pluralistic view.[26] In this schema, the process of economic planning is classified as pluralistic, because this expression is exactly applicable to it. Although the reports of the Economic Council often contain paternalistic expressions, they are just sermons and persuasions and do not pertain to decision-making on concrete policy measures.

Figure 1 is a diagram illustrating three patterns of government-interest group interaction. It is somewhat simplistic, but the intention is to provide a tool for distinguishing political and administrative events during the period under consideration. The three patterns may not be fully comparable in the same dimension, since they are brought together from the works of different authors. Still it is profitable to examine them to understand complicated events.

These events discussed above can be envisioned as a pendulum swing, starting from the centre, or the pluralistic process of economic planning. First, there was a swing toward the left with the Administrative Reform, then to the right with the first Maekawa reports, and finally back to the centre. It must be emphasized, however, that this motion was limited, never fully reaching either extreme. The essential character of government and the bureaucratic system in Japan, including advisory bodies, has remained pluralistic.

**Figure 1**

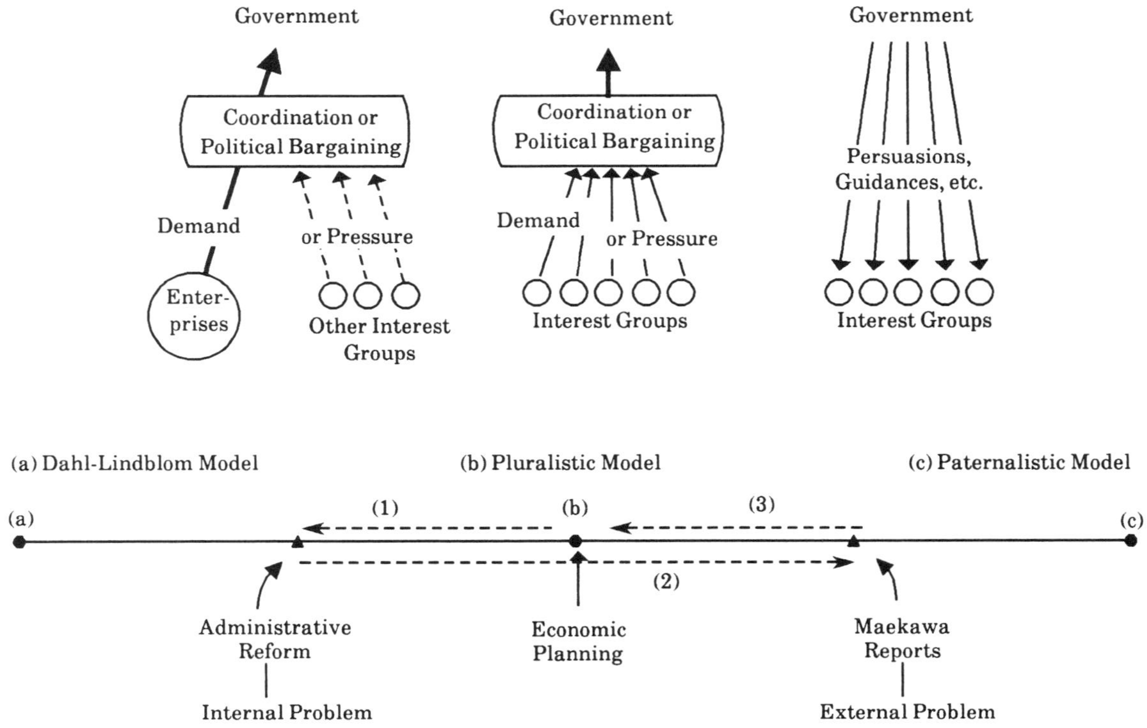

(a) Dahl-Lindblom Model

(b) Pluralistic Model

(c) Paternalistic Model

## Notes

1. Ryutaro Komiya, "Planning in Japan," *Economic Planning: East and West*, M. Bornstein, ed. (Cambridge: Ballinger, 1975).

2. Philip H. Trezise, with Yukio Suzuki, "Politics, Government, and Economic Growth in Japan," *Asia's New Giant*, Henry Rosovsky and Hugh Patrick, eds., (Washington: Brookings Institution, 1976), p. 790.

3. Examples of such sector specific plans are: The Long-Term Energy Supply-Demand Projection, The Basic Employment Policy Program, The Basic Plan for Electric Power Generation, The Long-Term Plan for Environmental Protection, The Medium-Term Defense Capacity Improvement Plan, etc. Medium-term programs promulgated by prefectural governments are also influenced directly by the national plan, or indirectly through The Comprehensive Regional Development Plan.

4. Trezise, p. 790-791.

5. John C. Campbell, "Contemporary Japanese Budget Politics" Ph.D. Dissertation, (Ann Arbor: University of Michigan, 1973).

6. Trezise, p. 792.

7. In that election, the ruling Liberal Democratic Party failed to recover the seats in the Lower House lost in the election of 1976. This near defeat was generally thought to be caused by the unpopularity of a new indirect tax.

8. Shouhei Shibata, "1980 Nendai Keizai Shakai no Tenbo to Shishin no Kihonteki Kangaekata [On Basic Views of 'Outlook and Guidelines for the Economy and Society in the 1980s']," *ESP*, (Tokyo: Economic Planning Agency, October 1983).

9. Ibid.

10. Note the frequency of the phrase "the vitality of the private sector." This echoes the emphasis placed on this concept in the reports of SCEAS. See Section 5.

11. Economic Planning Agency, "Outlook and Guidelines for the Economy and Society in the 1980s," (Tokyo: Government Printing Office, August 1983).

12. Ki-ichi Saeki, "Saikento site meikaku ni [Reconsider It and Make It Clear]," in "Comments and Impressions on the 'Outlook and Guidelines'," *ESP*, (Tokyo: Economic Planning Agency, October 1983).

13. Rincho-Gyokakushin OB Kai [Group of former members of the secretariat of the SCEAS], *Rincho to Gyokakushin* [The Special Committee and the Special Council], (Tokyo: Gyousei Kanri Kenkyu Centre, 1987), p. 4-6.

14. The crucial catalyst for the movement toward privatization and the dismembering of Japan National Railway (JNR) was actually this report from the Fourth Sub-Committee. Atsushi Kusano provides a very dramatic and exciting description of how some members of this Sub-Committee and their government sympathizers in the government and dissidents in JNR succeeded in publishing this appeal to the reform. See his *Kokutetsu Kaikaku* [Reforming JNR], (Tokyo: Chuo Koronsha, 1989).

15. Their proposals were adopted by one sub-committee and reflected for a while in the discussion of the main committee.

16. An ideal, but perhaps not feasible, solution would have been to both privatize the postal saving system and deregulate banking activities.

17. Admitting that the SCEAS succeeded in agitating public opinion for the need for administrative reforms, notably for the reform of JNR, Kusano concludes that the achievement of rather radical reforms owes very much to the personal abilities of particular committee, sub-committee and secretariat member to make persuasive arguments and to make appeals attractive enough to collect support from journalism.

18. I owe the observations in this paragraph and the preceding one to Osamu Nariai of the Economic Planning Agency. See his "Progress of Japan's Economic Restructuring and Future Tasks," (mimeo.), (Tokyo: Foreign Press Centre, 1988).

19. Ibid., p. 7.

20. The Economic Council, "Policy Recommendations of the Economic Council—Action for Economic Restructuring," (May 14, 1987), p.1. (English translation).

21. Ibid., p.6.

22. Note that the formal title of the first Maekawa group was "The Advisory Group on Economic Structural Adjustment for International Harmony."

23. "Japan's Viability in the Context of Harmony with the Rest of the World—Five Year Economic Management Plan" is the literal translation of the Japanese title. Somehow, the title of the English translation published by the Economic Planning Agency became "Economic Management within a Global Context".

24. These points are taken from David Vogel, "New Political Science of Corporate Power," *The Public Interest*, No. 87, (Spring 1987).

25. Masahiko Aoki, "The State and Markets: A Bargaining Game Theoretic Implication," *The State and the Private Enterprise in a Global Society*, ed. K. Uno, unpublished report prepared by the National Institute for Research Advancement, Tokyo, (September 1988).

26. For example, see M. Muramatu and M. Ito, *Chiho Gi-in no Kenkyu* [A Study of Representatives in Local Parliament], (Tokyo, Nihon Keizai, 1986), or T. Inoguchi, *Gendai Nihon Seiji Keizai no Kozu* [The Structures of Contemporary Politics and Economics in Japan], (Tokyo: Tokyo Keizai Shinposha, 1983). In the discussion above, the role of politicians and party politics in the Diet has been scarcely dealt with. There are some extensive studies of the role of "Zoku" (diet-tribes) and it may be a fascinating subject to examine if such a theory as presented by Takashi Inokuchi and Tomoaki Iwai can be applied to administrative reform in general or the reform of JNR in particular. Kusano in fact provides an interesting application of the Inokuchi-Iwai theory to the case of JNR reform, but he seems to rely there too much on the imagination of an outside observer. In addition, I agree with Kusano's observation that the activities of committee members and the secretariat, and support from journalism was effective enough to contain the influence of the diet-tribes that hoped to maintain the status quo. Thus I do not think it necessary to attempt an extensive analysis of the behaviour and influence of diet members on the process of administrative reform in this context. Interested readers are referred to Kusano, op. cit., and Inokuchi and Iwai, *Zoku-Gi-in no Kenkyu* [A Study of the Diet-Tribes], (Tokyo: Nihon Keizai Shinbunsha, 1987).

# Thailand

## Economic Policy-Making in a Liberal Technocratic Polity

*Chai-Anan Samudavanija*

The year 1980 is a watershed in modern Thai political-economy. It marks the beginning of a military-technocratic alliance which gradually gained momentum toward the mid-1980s and resulted in the institutionalization of a semi-democratic regime characterized by the depoliticization of the economic decision-making process. It is this grand coalition, under the leadership of General Prem Tinsulanond, which transformed Thailand from an import-substitution economy to an export-led economy.

The same period saw a rapid rise of a western-educated technocratic elite and the transformation of the bureaucratic polity. As the economy became more complex and dependent on a global capitalist system, it was no longer sufficient for a bureaucracy to engage in its traditional functions of the maintenance of law and order and legalistic control over the private sector.

Although Thailand has been experiencing a very rapid pace of socio-economic growth, its technocrats have always pursued a careful, conservative and incremental approach to development. The same attitude towards fiscal policy has been the standard operating procedure since the Second World War and has resulted in a stable currency.

Economic policy-making in Thailand has, for the most part, not been politicized. Perhaps this explains why Thailand has been able to achieve remarkable success in economic development compared with other developing countries. The depoliticization of the economic policy-making process occurred in the 1970s when the Communist Party of Thailand actively challenged the state and the capitalist system and the technocrats were able to convince the military leaders to leave economic decision-making to them as they were more "professional". While military factions fought among themselves, the technocrats and the academics were sought after as advisers to every faction and every government. Such continuity resulted in the stabilization and routinization of economic decisions.

This paper attempts to present an analysis of economic policy-making in what is termed "a liberal technocratic polity". I will first highlight patterns of socio-economic changes in Thailand and their impact on the bureaucracy. The paper will not deal in detail with cases in economic decision-making, but will present a general picture and a framework for a macro-analysis of the process.

Thailand's economy has grown rapidly over the past two decades, producing an average annual per capita income growth of almost 5 percent between 1960 and 1980. (In 1961, per capita income was Baht 2,137 compared with Baht 12,365 in 1980. US$1 = 22 Baht in 1980.) Over the same period, there has been a rapid transformation in the structure of production, with the share of agriculture in total value added declining from 40 percent in 1960 to 25 percent in 1980.

The manufacturing sector expanded rapidly as a result of import substitution. Its share of GDP rose from 10.5 percent in 1960 to 18 percent in 1980. The number of factories increased fivefold during the same period. In 1980 there were 3.6 million workers in the industrial and service sectors. In addition to the growth of the labour force employed in privately-owned factories, there was also a rapid increase in the number of workers in state enterprises, from 137,437 in 1973 to 433,649 in 1983.

Labour unions in state enterprises have been more politically active than labour unions in the private sector. In 1983, there were 323 labour unions in the private sector and 91 state enterprise labour unions. However, the former had only 81,465 members compared with 136,335 members in the latter. Public enterprise workers in the electricity, rail and water supply industries are the most organized; their political significance due largely to their control of public utility services in metropolitan areas, which gives them considerable bargaining power.

Socio-economic change in Thailand has been urban dominated, with major potent political forces concentrated in the capital city. After two decades of development, Bangkok retains a remarkable degree of primacy. While only about 9.7 percent of the Thai population lived in Bangkok in 1980, the city accounted for 32.7 percent of total GDP. Although the overall incidence of poverty was reduced from 57 percent in the early 1960s to about 31 percent in the mid-1970s, poverty remains largely a rural phenomenon.[1] It is estimated that in 1980, 11 million people in the rural areas lived in poverty. The benefits of growth have not been evenly dispersed. The gap between the rich and the poor, and between the rural and urban sectors, has widened.

By far the most important change in the Thai economy since the 1960s has been the rapid expansion of large-scale enterprises (those with assets of more than Baht 500 million). According to one study,[2] the value of capital owned by these enterprises amounted to nearly 74 percent of GNP in 1983.

This growth of semi-monopolistic capital was encouraged by government development policies during the authoritarian regimes of the late 1950s and 1960s which favoured the development of industrial capital over agriculture. Such policies were aimed at transforming the agricultural surplus into a manufacturing base. Policies of import substitution and trade protection were implemented. During the same period, government after government pursued a policy of price controls which favoured urban communities at the expense of the agricultural workforce.

In sum, economic development in the past two decades has resulted in the concentration of economic power in the capital and has created a large urban working class. At the same time, this development was paralleled by the growth of the bureaucracy, which, while remaining highly centralized, penetrated more into the rural areas. By 1980, the number of government employees (excluding the military) reached 1.4 million, making the ratio between population (46 million) and government employees, 33 to 1. In the same year, government expenditure on personnel services accounted for 35 percent of total government expenditures.[3]

During the last three decades the Thai economy has also become increasingly linked with the global economic system. In the 1970s export growth rates substantially exceeded GDP growth rates, and exports now include a very high proportion of manufactured products. The country's links with the global capital system have also increased dramatically in the area of foreign investment. Foreign investors are now involved in enterprises totalling over Baht 23 billion in registered capital, which is distributed over a range of key sectors, most notably oil and gas exploration and production, commerce, services, construction and manufacturing.[4]

The direction of Thailand's development has had a number of socio-economic consequences relevant to the discussion at hand. First, the last three decades have witnessed a doubling of Thailand's population to its current level of 56 million. At the same time rapid urbanization has taken place. In 1960, just over 10 percent of Thais lived in urban areas, but by 1983 this percentage had increased to nearly 25 percent. The availability of educational services has also risen fast, with the percentage of Thailand's population enrolled in higher education rising from 2 percent in 1960 to nearly 15 percent in the early 1980s.[5] According to one estimate, by 1991 there will be one million students in universities.[6]

Second, there has been a rapid expansion of the middle class, mainly in, but not confined to, Bangkok. Numbering only 178,000 in 1960 and 284,000 in 1970, Bangkok's middle class increased to 1,800,000 in 1986 or the equivalent of 31 percent of the city's population. A similar, but less dramatic trend, has also taken place in other urban centres. The provincial middle class constituted some 19 percent of the provincial urban population in 1986. The middle class now represents 7.5 percent of total employed persons, while the working class and farmers make up 23.5 percent and 57.6 percent respectively.[7] Although obviously a relatively heterogeneous group, those belonging to the "new" Thai middle class share a number of common characteristics. They are young (mostly in the 25-35 age group); highly educated (bachelor's degrees); have a small family with a working spouse and a two-bedroom house in either government or private housing estates;

like to travel, read, tune in to radio and television programs, especially those on current affairs; and, if they do not have them already, aspire to own a car and possess a credit card.[8]

It was clear by the early 1970s that the process of economic growth and development which had taken place during the previous two decades had brought profound changes. Rapid expansion of the urban-industrial sector and of the educational system fostered greater political pluralism, political awareness and ultimately political mobilization when it become obvious that the military regime would continue to resist the demands for greater political participation and, if necessary, to continue to rule by force. The October 1973 student-led uprising against the regime finally brought to an end the military's complete domination of Thai politics. Between 1973 and 1976 the flower of democracy bloomed, with the Thais' political awareness and mobilization reaching heights never before or after attained.[9] Reactionary coups in 1976 and 1977 cut short the era of open politics, but evidently the so-called 14 October 1973 Revolution, and the experiences of political participation made possible by it, had a deep and long-term impact upon the structure and processes of Thai politics. The military could no longer hold the monopoly of power and prevent the existence of participatory political institutions which, in turn, have begun to develop a life and a momentum of their own.

It is indisputable that socio-economic changes led to the emergence of new groups in society, but whether the existence of these groups will lead to a pluralist democracy is another matter. In the case of Thailand, socio-economic changes occurred under situations of semi-imposed development. In this pattern of development, political and administrative structures such as the military and the bureaucracy have been able to grow alongside the growth of the private sector. In fact, they have been able to create new institutional structures of their own or to adjust existing structures and functions to cope with pressures coming from extra-bureaucratic forces. Hence, changes resulting from social and economic modernization have not automatically strengthened voluntary associations and political groups because the transformation of economic power into political power has been delayed by the relative autonomy of the state in policy-making matters. In other words, socio-economic changes in Thailand have enabled the non-bureaucratic groups to participate more in bureaucratic politics rather than to fundamentally change the nature of the Thai political system from that of a "bureaucratic polity" to that of a "bourgeois polity".

In recent years economic development has brought increased criticism of the bureaucratic polity and of military domination of politics. Ansil Ramsey has observed that political participation in decision-making in Thailand has recently extended to "bourgeois-middle class groups", especially the business elite, who have begun to play a major role in Thai cabinets and in economic decision-making. Other groups from middle class backgrounds, such as leading academics and technocrats, also have increased their access to decision-making.[10]

But it is simplistic to conclude that the bureaucratic polity has evolved into a bourgeois polity, or is developing in that direction. Perhaps it is safer to say that the bureaucratic polity has been transformed into a technocratic polity; an argument which will be pursued further in the case studies of the Joint Public and Private Consultative Committee (JPPCC) and the Council of Economic Ministers (CEM) below.

The interplay between the various state and non-state institutions involved in the economic policy-making process in a pluralist democracy, an authoritarian bureaucratic polity, and a liberal technocratic state takes different forms. In the case of Thailand, rapid socio-economic changes posed a great challenge to the traditional bureaucratic polity which quickly found that existing administrative structures were inadequate to deal with exogenous forces. As the planners admitted:

> The state's administration of the economy has emphasized the solving of immediate problems in various fields, without laying any foundation for social and economic structural adjustments in order to solve long-term problems. In particular, there has been no administrative development reform both at the central and local levels to be in tune with changes or to adjust the economic administration system to catch up with economic and fiscal situations as well as world politics which have undergone tremendous changes.[11]

The Fifth National Economic and Social Development Plan (1981-1986) recognized the urgent need for more efficient economic management in a country where socio-economic changes had been very rapid in the previous two decades. While the Thai economy during 1960-1980 experienced a dramatic structural change,[12] its administrative system was basically a bureaucratic-authoritarian entity which was highly centralized and designed to "administer" rather than "manage". There was a great reliance on rules, regulations, and clumsy procedures, thus causing unnecessary delays and bottle-necks in the process of government administration, particularly in the area of international trade. As the Thai economy became more integrated with the world economy, there arose an urgent need to evaluate and redefine the role of the state and its public sector vis-à-vis that of the growing private sector. The opening up of the Thai economy also meant that it had to be more enterprising, competitive, and efficient. There was also a great need for improved coordination and cooperation between the public and the private sectors.

In the past, the financial-business community had suffered from a number of handicaps. From the earliest days of development of the capitalist system in Thailand, this community was dominated by ethnic Chinese, who in turn were dependent upon, and controlled by, first princes and nobles during the absolute monarchy period, and then by civilian and military bureaucrats after 1932. This position of internal dependence meant that the financial-business community was vulnerable to the government's nationalistic campaigns and to government leaders' pursuit of self-interest. Indeed, the years between 1947 and 1973 were most notable for the stranglehold that military leaders had on this community. It was this period of "Commercial Soldiers", when different military factions built up their economic bases by setting up their own companies, securing control over state enterprises and semi-government companies, and gaining free shares from Chinese-owned private firms. They subsequently used these economic bases to extend further their influence within the body politic.[13]

The process of economic growth in Thailand helped to transform this quasi-colonial relationship between the civilian-military bureaucracy and the Chinese financial-business community; a relationship that over the long run was becoming less unequal in any case due to inter-marriages and the two elites' increasingly similar socialization processes, particularly where education was concerned.

Due to the vast expansion of the industrial, agro-industrial, trading, financial and non-financial services sectors and also to the rapid increase in the linkages with the outside world, the financial-business community has been transformed from a basically dependent entity into a more independent and organized sector; innovative and liberal in its outlook and dynamic and hardworking in the conduct of its affairs. This transformation is reflected in the present political position and role of political parties, with whom the financial-business community enjoys an increasingly close—some would say symbiotic—relationship, and also in the fact that the government is paying greater attention to the private sector and its needs. For example, recently 80 major construction companies successfully pressured the government to reduce import taxes on steel bars and billets in order to compensate for losses arising from sharply increased construction material costs.[14] The government's greater sensitivity to the demands and needs of the local private sector has also extended to the foreign private sector involved in investing in and trading with Thailand. This was demonstrated recently when the government agreed to re-examine the Port Authority of Thailand's policy regarding the use of cranage at Bangkok's busy Klong Toei Port in response to both local and foreign firms' calls for a more equitable and efficient arrangement.[15]

However, this process of "liberalization" in the economic arena also has its limits. In some industrializing countries the expansion of the middle class, the "bourgeoisie", or the entrepreneurial class, may serve as a force of democracy if such classes can forge an alliance with industrialists and others to control the state, democratize it and put it in the service of economic development.[16] In Thailand this has not been the case, for the Thai bureaucracy has been able to transform itself during the process of economic growth by actively promoting economic development and coopting first the technocrats, and then the private sector.

In the late 1950s and early 1960s, Thai military leaders incorporated economic development as an integral part of their strategy to create stability and security for both the regime and the nation. With the assistance of foreign and local advisers, the government promoted what can be termed the "ideology" of development, which helped to transform a faction-ridden bureaucratic polity into a more unified activist bureaucratic state. The Budget Act of 1959 established a central budget office directly responsible to the chief executive. A Budget Procedures Act of the same year gave an appreciable degree of order and coherence to the budgetary process. These reforms in the legal and organizational bases of public financial administration, together with the inauguration of the First National Economic Development Plan in 1961, resulted in the subsequent rise to power of a technocratic elite which gradually became the power base of a new military-technocratic alliance. The activist role of the state can be seen in the growth of the administrative structure and changes in the pattern of government spending in the period after 1957. In 1957 there were only 90 departments and 550 divisions; in 1979 there were a total of 131 departments and 1,264 divisions.[17]

This qualitative and quantitative development of the bureaucracy made it possible for the government to allow the financial-business community to participate in some areas of the decision-making process under terms of reference determined by the state at a time that the private sector was rapidly expanding. This process of cooption has meant that the latter does not present a real challenge to the bureaucratic-activist, technocratic state but strengthens it because the state's decision-making capacity is enhanced by the private sector's participation as a source of information, expertise and, ultimately, legitimacy. Moreover, once the private sector is given this channel, the *need* for the financial-business community to seek alliances with political parties in order to influence policy decisions is lessened. Although obviously many financiers and businessmen "go into politics" to try to exert control over the nucleus of state power, many more do not, preferring to deal directly with the military and technocrats. For them access to, rather than possession of, political power is a precondition for the accumulation of wealth. In this sense the relationship between politics and economics is essentially one of exchange, rather than one involving the transfer of power from the economy to the polity.[18]

The impetus toward cooption of the private sector stemmed from the domestic and international political situation in the mid-to-late 1970s. This period contained a series of traumatic experiences for Thailand, creating fear and uncertainly among the upper class, the civilian and military bureaucrats, and the financial and business community. These experiences included the student revolution of October 1973, followed by a lengthy spell of student activism; the fall of Indochina to communist forces in 1975 and the U.S. military withdrawal from mainland Southeast Asia in 1975-76; the violent right-wing coup of October 1976; border problems with Khmer Rouge Cambodia in 1977-78; and Vietnam's invasion of Cambodia in 1978-79. It was during this period that those in charge of Thailand's security policies came to the conclusion that security was a multi-dimensional concept and that real security involved not only self-reliance in defence, but also a balance among military, political and socio-economic requirements, with priority to be given to economic development.[19] This line of thinking was pervasive among the upper echelons of Thai policy-makers and it was during this period that Dr. Snoh Unakul, one of the top technocrats in the government circle, developed his strategy of public-private sector cooperation, based on the premise that if Thailand was to survive and prosper, "progressive captialists" should be encouraged to join hands with the government and "exploitative capitalists" should be kept under control in order to avoid the disarray which political chaos could bring to the market economy.[20]

Significantly, the Fifth National Economic and Social Development Plan (1982-86), drafted by civilian and military bureaucrats in close collaboration with one another, identified as the main objective of Thailand's development strategy the achievement of "national stability", which was to be brought about by the balancing of security and socio-economic requirements and according priority to economic development.[21] The Fifth Plan also devoted a great deal of attention to the issue of public-private sector cooperation. Chapter 6 of the Plan was entitled "The Mobilization of Cooperation from the Private Sector"[22] and under the section on "Problems" it stated, with uncharacteristic bluntness, that the relationship between the two sectors was very poor:

> The main problem of the public sector is the inefficiency of the bureaucracy, which makes laws and regulations

obstructive to development, while the private sector which has to comply with those laws and regulations, possesses more efficiency in management, but seeks to make profits without being responsible to social justice in society as a whole.[23]

It was pointed out that the private sector should be encouraged to improve its management capacity in order to respond to increased responsibilities.[24]

At the same time, the Fifth Plan pointed out that there was an urgent need to study and research the pattern and form of the existing institutions of the private sector in order to develop them into effective instruments which could serve the state, and to find ways and means of facilitating this process.[25] It also went on to propose a system and a mechanism for public-private sector cooperation which was to take the form of a high-level joint committee.[26] Clearly, the new wave of thinking was to liberalize and privatize, with the technocratic state shifting its role from that of control and containment to that of supervision and promotion. As the Fifth Plan stipulated, the state would "delegate responsibility and some powers to the organizations (of the private sector) in order to lessen the burden of the state".[27]

Accordingly, the Joint Public and Private Consultative Committee (JPPCC) was formed, with the Prime Minister as Chairman and the NESDB Secretary-General (Dr. Snoh) as Committee Secretary. Other members were deputy prime ministers, ministers and deputy ministers of economic ministries, the Governor of the Bank of Thailand and the Secretary-General of the Board of Investment. The private sector was represented by three representatives from the Thai Chamber of Commerce, three from the Association of Thai Industries and three from the Thai Bankers Association. The structure of the JPPCC is shown in Figure 1.

The agenda for JPPCC meetings was decided by the Joint Standing Committee on Commerce, Industry and Banking, where medium-sized and small business concerns are poorly represented. These groups thus preferred to by-pass the JPPCC structure altogether and develop direct contacts with the bureaucracy.

The creation of the JPPCC at the national level was followed by development of provincial JPPCCs.

**Figure 1**
**The Joint Public and Private Consultative Committee (JPPCC)**

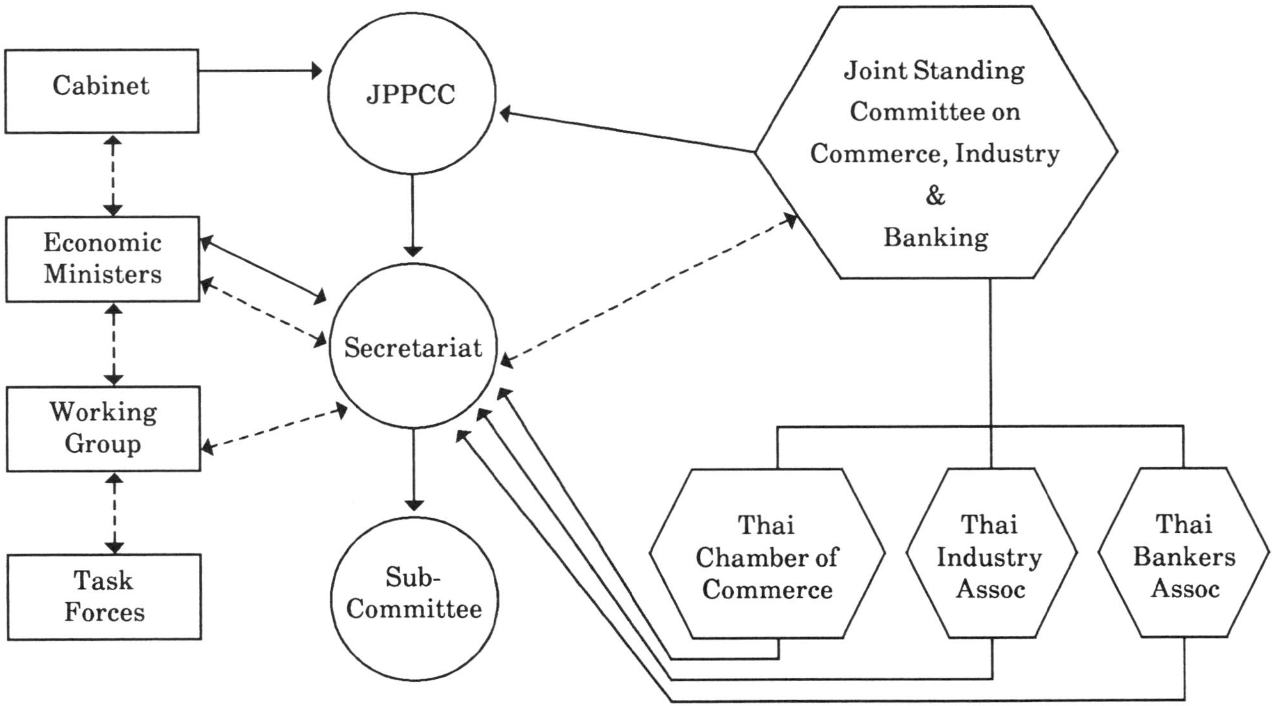

In 1967 there were only two chambers of commerce and from 1968 to 1976 no new ones were established because the government feared their potential political influence. From 1980 to 1983 only eleven provincial chambers of commerce were registered. After the government-sponsored meeting of regional JPPCCs in Chiengmai in February 1984, 23 chambers of commerce were established in various provinces. From 1985-1987, another 34 provincial chambers of commerce were established with encouragement and guidance of the government. The administration of provincial JPPCCs is under the office of the governor who is the chairman of the committee. It is natural that the dynamism, or the lack thereof, of these bodies in determined by the governors.

The structure of the JPPCC both at the national and provincial levels is basically bureaucratic, with officials outnumbering representatives of the private sector. In a series of interviews[28] conducted by the author, representatives of the private sector all agreed in principle that the set-up was useful, but some were critical of its effectiveness and thought that the JPPCC was merely a "talking stadium" (one even said that it was a "do-nothing" body), while others were of the opinion that having such a body was a psychological breakthrough for public-private sector relations, which used to be quite antagonistic. Another said the national JPPCC was not a free and open platform but was overly guided by the Secretariat.

It is evident that the JPPCC is a mechanism created and utilized by technocrats to arrange relations between the state and the private sector. It is a vehicle used by the technocrats, not only to establish a constructive alliance with industrialists, bankers, and businessmen, but also to put pressure on the bureaucracy to initiate policy reforms.[29] In this sense, the technocrats are taking the role of reformers as well as guardians of the "national interest". The JPPCC has established a standard operating procedure, barring matters which are of interest only to specific firms and accepting only matters concerning the general interests of a sector or a type of industry.

The establishment of the JPPCC and the keen interest taken in it by the top government leadership can be regarded as an important safety valve which helps reduce conflict and tension between the military, technocratic-bureaucratic elites and the financial-business community. Its existence presents an alternative for the emerging economic forces not to join or set up political parties but to use the JPPCC as a channel of communications with the top leadership. Professionals in the three associations prefer to go through the JPPCC and are committed to develop it rather than to travel along the political route. The creation of the JPPCC can therefore be regarded as a non-political body which strives to reach consensus and work together in the spirit of "krengjai" (deference), instead of going through the process of competition and bargaining in political parties and parliament.

The basic problem of the JPPCC is its composition, which has been criticized as representing only big business interests. There has been a great concern to promote provincial chambers of commerce in order to strengthen the economic base of the Thai capitalist system. The Center For International Private Enterprise (CIPE) of the American Chamber of Commerce has been active in funding training for provincial businessmen since 1983. CIPE has granted $97,000 for a six-year program aimed at training all members of the 73 provincial chambers of commerce.

It seems that there is a conscious effort to support the development of a strong private sector throughout the country. The emphasis is on the provinces not the centre and, in fact, the policy is to assist provincial private enterprises to be independent. This strategy of economic pluralism is an important mechanism in creating stable and independent economic and social forces outside the primary city. Provincial chambers of commerce, once created, could become "instruments of the state" at least in supplying information and articulating and aggregating demands. These structures are very political in nature but are nevertheless regarded as non-political. Whatever the original intention may have been, the JPPCC and provincial chambers of commerce are state-led pressure groups which have made political parties dysfunctional as participatory institutions.

Economic policy-making in Thailand has always been an arena for bureaucratic competition, accommodation and mutual adjustment rather than for participation by the mainstream political process. The initiation of a National Economic Development Plan and the establishment of the National Economic and Social Development Board and other central policy agencies in the 1960s were instrumental in strengthening the role of the state vis-à-vis non-state institutions. It is important to note that central government policy agencies have been able to function mostly under a strong non-elected executive and have until recently always enjoyed a high degree of freedom from political control by elected representatives.

There are nine core agencies which have a direct role and influence in policy formulation and resource allocation for national development. They are the office of the National Economic and Social Development Board (NESDB), The Bureau of the Budget (BOB), the Fiscal Policy Office (FPO) of the Ministry of Finance, the Office of the Board of Investment (BOI), the Office of the Civil Service Commission (CSC), the Comptroller General's Department (CGD), the Office of the Auditor General (OAG), the Office of the National Education Commission (NEC), and the Bank of Thailand. These core agencies are given logistic support by six additional agencies, which include the Secretariat of the Prime Minister, the Secretariat of the Cabinet, the Department of Technical and Economic Cooperation, The National Statistical Office, the Office of National Energy Administration, and the National Environment Board.[30]

Among these core agencies, the most significant are the NESDB, the BOB, and the FPO. They are charged with planning, administration, evaluation and follow-up of development project analysis and resource allocation for development work; analysis of the operation of government agencies; and analysis of government revenue (as well as estimating revenue) and expenditures and tax collection.[31]

This paper will not attempt to describe in detail the roles and functions performed by these agencies. More important is an explanation of the significance of their roles and functions in relation to the top policy-making institution—the cabinet. In a country where active participant political institutions were first undertaken by the state, these technical core agencies became the state's major mechanisms not only for planning but also as instruments of control over the bureaucracy.

The NESDB, the BOB, and the FPO are in strategic positions to set parameters for cabinet-level policy decisions. They also set parameters for bureaucratic competition for resources as they screen all government agencies' proposed development projects.

As the Thai economy has become more diversified and interdependent with the global economy, the roles of these core agencies have been enhanced by the need for the more sophisticated technical inputs and technocratic skills which military leaders usually lack, resulting in a gradual increase in influence and power of the technocrats in the policy-making process. What development has brought, therefore, is not only the emergence of new socio-economic forces in the civil society, but also a qualitative transformation in that part of the bureaucratic polity which is the technocratic aspect of the system.

While military factions fought each other and took turns in "ruling" the country, it has been the technocrats who have actually steered the ship of state. It is well known that all of the policy statements issued by governments since 1961—be they elected or non-elected—have always been "Xerox copies" of national development plans or short versions of these documents.

It would, therefore, be simplistic to equate economic pluralism with pluralist democracy which assumes an active and wide participation of interests and pressure groups in the policy-making process. It is important to look more closely at the *structure* of that process and the degree to which the structure allows for participation.

The success of Thailand in its economic performance during the past three decades is the *flexibility* and *pragmatism* of the "state within the state". The semi-democratic arrangements of the Prem leadership is a reflection of this flexibility and pragmatism. It was under this long and uninterrupted period (1980-1988) that an authoritarian-bureaucratic system was transformed into a liberal technocratic mechanism of the state.

The need for such an adjustment was increasingly recognized during the 1970s, particularly after the 1973 student revolution and its aftermath.[32] Thai society in the late 1970s was significantly different from that of the early 1960s when Field Marshal Sarit could so proudly and confidently proclaim that he was solely responsible for the administration and governance of the country.

The period of semi-democracy (1980-1988) coincided with the Fifth and Sixth Plans (1982-1986; 1987-1991). Under a semi-democratic system, political parties and parliament are "additional" institutions in the economic policy-making process. How could the military and their technocratic allies cope with this changing situation? How could they steer the course of national development through the rapidly changing international environment? How could they safe-guard their power and influence vis-à-vis other pressure groups? How could they prevent political parties and pressure groups from asserting demands which they considered detrimental to the national interest and development?

The military and technocratic elite alliance under the leadership of General Prem Tinsulanond developed a number of mechanisms to cope with the aforementioned problems, the most important of which was the establishment of an Council of Economic Ministers (CEM).[33]

The CEM is not a cabinet committee or an inner-cabinet in the tradition of the Westminster model, but a committee composed of three elements, namely, cabinet ministers who represent government coalition parties, cabinet ministers who were the Prime Minister's appointees (non-party members), and the Prime Minister's advisors and chief technocrats.

The Council of Economic Ministers evolved from the Economic Policy Committee (EPC) which was established in the first Prem government in early 1980. The EPC was intended to be an inner cabinet designed to supervise, analyse, evaluate, and determine the direction of economic policies. It was composed of sixteen members chaired by the Prime Minister. The other members were 4 deputy prime ministers; the ministers of Finance, Agriculture and Cooperatives, Industry, Communications, and Commerce; the Governor of the Bank of Thailand; the Secretary-General of the Secretariat of the Prime Minister; the Director of the Budget Bureau; the Chairman of the Prime Minister's Advisors, Dr. Phaichitr Uathavikul; and the Secretary-General of the NESDB as the secretary of the committee. The EPC was, therefore, not a sub-structure of the cabinet, but independent from it. Although it was not part of the cabinet, its decisions had the power of cabinet decisions. Its power was enhanced by the leadership of the Prime Minister who, under the advice of the advisors and the NESDB, referred all major important policy matters proposed by cabinet ministers representing coalition parties of other ministries to the EPC.

Apart from this top-level decision-making structure, the Prime Minister appointed a number of national committees covering all important policy areas such as the Energy Committee, the Eastern Seaboard Committee, the Rural Development Committee, and the Administrative Reform Committee. These committees were either chaired by the Prime Minister himself or by senior ministers who were the Prime Minister's appointees. In every committee, academics who were advisors to the Prime Minister and technocrats from NESDB, BOB, and FPO played active and influential roles in the deliberations of the committees.

In the first two years of General Prem's leadership, intense conflicts developed between the coalition government's members, resulting in a cabinet reshuffle and an attempted coup in April 1981. After that, the Prime Minister, his advisors, and the technocrats gradually tightened their power in every aspect of the policy-making process with the assistance of the secretariat of the NESDB which, by the mid-1980s, had become the most powerful core agency, and, understandably, the target of attack not only from political parties but also from other government agencies.

In the second Prem government (1981-1983), the Council of Economic Ministers replaced the EPC. This time the Council became larger, comprised of 25 members, and in 1986 the members increased to 30. In 1980, there were only 5 full ministers on the EPC; in 1983, there were 10 ministers and 2 deputy ministers representing 8 out of the 13 ministries; and in 1986, there were 11 ministers and 3 deputies representing 9 ministries.

Although political parties in the coalition government gained their ministerial portfolios according to their strength in the parliament, in the CEM, they were allotted equal representation on the committee (3 each from the 3 political parties in 1983, and 4 each from the 3 political parties in 1986). The rest were ministers who were the Prime Minister's appointees (non-party cabinet members), the Prime Minister's advisors, and chief technocrats. In the 25-member CEM in 1983, only 10 came from political parties while 8 were non-party cabinet members and 7 were advisors and technocrats. In 1986, only 13 out of 30 members of the CEM were party members.

It was not only the composition of the CEM which reflected a conscious effort to depoliticize the decision-making structure. The operating procedures of the CEM were also designed to make deliberations of the council, essentially technocratic. Committee decisions were made by consensus, not by voting or by resolutions, and the Prime Minister relied heavily on the advice and recommendations of his advisors and chief technocrats, especially Dr. Snoh Unakul, Dr. Phaichitr Uathavikul and Dr. Virabongse Ramanagkura.

As the top decision-making structure, the CEM made it possible for a new source of power to exert itself as a countervailing force to political power coming from participant political institutions. This is the "referent power" of the academics and technocrats. It has been this type of power which stabilized and legitimatized the semi-democratic regime for nearly a decade.

The secretariat of the NESDB which served as the secretariat for the Council of Economic Ministers, the Eastern Seaboard Committee, the National Rural Development Committee, the Energy Policy Committee, and the Joint Public and Private Consultative Committee, was, therefore, at the center of a network spread across major policy areas. Such a pattern of decision-making process was a conscious effort in depoliticization which was greatly needed under the period of redemocratization in the late 1970s and throughout the 1980s.

Important and difficult policy decisions such as devaluation of the baht, the implementation of an austere fiscal and monetary policy, a conscious rural development effort, and the prevention of projects motivated by vested interests or partisan political considerations were made possible under such arrangements.

The CEM is, therefore, the technocratic response to a rapidly changing socio-economic and political situation, while the JPPCC represents an attempt by the state to maintain and exert its autonomy towards big business groups in society. The JPPCC can be regarded as an alternative for pressure groups to participate and influence policy decisions without identifying and allying themselves too closely with political parties. The consultation between state and non-state agencies in Thailand, therefore, revolves not around participant political institutions such as political parties and parliament but around a unique kind of arrangement which can be called political gradualism and the technocratization of economic policy-making. By the late 1980s such structures, sub-structures and processes have become so deeply institutionalized that when the new cabinet under the first prime minister to be an elected member of parliament in a decade and a half was announced in September 1988, the mass media and influential members of the private sector, the academia, and the general public simultaneously expressed their concern and scepticism about the ability, integrity, and effectiveness of the government composed almost entirely of elected politicians. It remains to be seen whether a full-fledged democratic system could continue to utilize and gain the respect and support of the technocrats in the same way that the semi-democratic regime did. In other words, the question of whether the process of democratization will contribute to economic development or hinder it remains unresolved. But one thing is sure. The technocrats have to adjust and redefine their role in this rapidly changing internal political situation. They have done it well in the past, so there is no reason why they will not be able to do it again.

## Notes

1. See details in *Thailand : Managing Public Resources for Structural Adjustment*, (A World Bank Country Study, Washington D.C.: The World Bank, 1984), pp. 1-13.

2. Krirkiat Phipatseritham, "The World of Finance: The Push and Pull of Politics", paper presented for the seminar on "National Development of Thailand Economic Rationality and Political Feasibility", Thammasat University, Bangkok, 6-7 September 1983, p.21.

3. Chai-Anan Samudavanija, "Introduction", in *Report of the Ad Hoc Committee to Study Major Problems of the Thai Administrative System*, (Bangkok: Administrative Reform Committee, 1980), pp. 6-7.

4. For recent figures on foreign investment, see "Foreign Investments in Thai Firms", *Bangkok Post*, 13 February 1987, p. 19, and 1987 *Year-End Economic Review*, op. cit., pp. 17-18.

5. According to figures provided by the National Economic and Social Development Board.

6. *Bangkok Post*, 13 February 1987, p. 3.

7. According to figures provided by Labour Research Branch, Labour Studies and Planning Division, Department of Labour, Ministry of Interior, Thailand.

8. *Deemar Media Index*, 1986.

9. Perhaps the most comprehensive treatment of Thai politics in the 1973-76 period is provided in David Morell and Chai-Anan Samudavanija, *Political Conflict in Thailand: Reform, Reaction, Revolution*, (Cambridge, Ma.: Oelgeschlager, Gunn & Hain Publishers Inc., 1981).

10. Ansil Ramsey "Thai Domestic Politics and Foreign Policy", paper delivered at the Third U.S.–ASEAN Conference, Chiengmai, Thailand, January 7-11, 1985, p. 4.

11. The Fifth National Economic and Social Development Plan, Chapter 8, p. 396.

12. The share of agricultural products in the country's total value-added dropped dramatically from some 40 percent in 1960 to just over 23 percent in 1985. The manufacturing sector saw its share of value-added rise from 12 percent in 1960 to 21 percent in 1985.

13. Chai-Anan Samudavanija, *The Thai Young Turks*, (Singapore: Institute of Southeast Asian Studies, 1982), pp. 14-22.

14. *Bangkok Post*, 15 June 1988, pp. 15, 30.

15. See *Phu Jad Karn*, 16-22 May 1988, p. 1,7 and 8; and *The Nation*, 16 May 1988, p. 19; and 17 May 1988, p. 19.

16. See Fernando Cardoso, "Entrepreneurs and the Transition Process: The Brazilian Case", in Guillermo O'Donnell, Philippe C. Schmitter and Laurence Whitehead eds., *Transition from Authoritarian Rule*, (Baltimore: The Johns Hopkins University Press, 1988).

17. Woradej Chantasorn, "The Expansion of Agencies in Thai Bureaucracy", paper prepared for the seminar on "Thailand in the 1980s", Pattaya, Thailand, 13-16 December 1979, p. 4.

18. Chai-Anan Samudavanija, "Macro-economic Decision-making and Management of Economic Policies", paper presented at the Conference on "Who Determines Thai Economic Policies?", Thammasat University, Bangkok, February 1988.

19. See a more extensive discussion in Sukhumbhand Paribatra, "Thailand : Defence Spending and Threat Perception", in Chin Kin Wah ed., *Defence Spending in Southeast Asia*, (Singapore: Institute of Southeast Asian Studies, 1987), pp. 84-90.

20. Author's interview with Dr. Snoh Unakul. Dr. Snoh was formerly Governor of the Bank of Thailand and is presently Secretary-General of the NESDB.

21. Paribatra, "Thailand: Defence Spending and Threat Perception", pp. 84-90.

22. NESDB, *Fifth National Economic and Social Development Plan, B.E. 2525-2529*, (Bangkok: NESDB, Office of the Prime Minister, 1982), pp. 417-419.

23. Ibid., p. 418.

24. Ibid., p. 419.

25. Ibid., p. 418.

26. Ibid., p. 419.

27. Ibid., p. 418.

28. March-May 1988, in Bangkok, Songkla and Chiangmai. The subjects of the interviews were 10 people who sat on the JPPCC at the national and provincial levels.

29. Prominent among these are tax system reforms and deregulation of public utilities.

30. TURA, "Thailand's National Development: Policy Issues and Challenges", (Bangkok: TURA-CIDA, August 1987), p.44.

31. Ibid., see details on legal mandate, roles and functions of these agencies, pp. 45-64.

32. For details see Morell and Chai-Anan, *Political Conflict in Thailand*.

33. For details on the Economic Ministers Committee see Chai-Anan Samudavanija, *Economic Ministers Committee*, (Bangkok: Thailand Development Research Institute, August 1988).

# South Korea

## Political Transition and Economic Policy-Making in South Korea

*Chung-Si Ahn*

## 1. Introduction

South Korea is a small, over-populated country with a poor resource base. When it became independent in 1948, South Korea was split with the North, losing its important economic base for natural resources and industry. The country was war-torn during 1950-53, severely damaging its infrastructure and remaining industrial capability. The 1961 coup by Park Chung-Hee followed 12 years of authoritarian rule by Rhee Syung-Man. Park launched ambitious economic development programs in the mid-1960s, making economic growth the nation's foremost priority task. He also used industrialization as a means for enhancing the political legitimacy of his authoritarian government. Since the nation's first five year plan in 1962, Korea has experienced almost miraculous economic growth and social transformation.

The South Korean economy has expanded at an average annual rate of nearly 10 percent in real GNP growth for the last two decades. This growth record has been matched by few other countries in the world. Even during the politically tumultuous period of 1983-87 under former president Chun Doo-Hwan, the average annual growth rate of the economy was, at 10 percent, the world's fastest—compared to 9 percent in Taiwan (4th), 4.7 percent in Singapore (12th), and Japan's 3.6 percent (18th). In 1987, the South Korean economy grew at a rate of 12.0 percent. Its GNP amounted to $118.6 billion, while the GNP per capita was $2,826. The country ranked 7th in the world in trade surplus, 8th in domestic savings rate, 12th in total trade volumes, 15th in automotive production, 18th in its size of GNP, and 24th in per capita telephone installment. The total trade volume of the country amounted to $88.3 billion, of which $47.28 billion were exports and $41.02 billion imports. In 1988 the economy continued to grow (12.1 percent), reaching a per capita GNP of $3,728. Total exports jumped to $60 billion, and imports to $52.5 billion. The trade surplus reached a record $13.8 billion. It is expected that the country's current foreign debts of $32 billion will go down to $28.5 billion by the end of 1989, and net foreign assets will increase from $25.3 billion in 1988 to $31.5 billion before the end of 1989.

Compared to its industrial capability, Korea's domestic market is relatively small. This makes its economy heavily dependent on external markets. For example, export earnings were equivalent to 40 percent of Korea's GNP in 1987 and four million workers are employed in the export industries. At the same time, Korea's export markets are concentrated in a few industrially developed countries. Korea is the 6th largest trading partner to Canada, with $0.88 billion imports from Canada and $1.39 billion exports; it is the 8th largest exporter to the 12 European Community (EC) members, earning 7.1 billion U.S. dollars from her exports to the EC; and Korea depends on the Japanese market for 17.8 percent of her total export earnings. The United States, however, is by far the most important market for Korean exports. In 1987, 38.7 percent of Korea's export earnings came from trade with the United States. This included 42 percent of the export earnings from electronic products, 80 percent of automobiles, 66 percent of footwear, and 47 percent of machine tools. Furthermore, this trade dependence with the United States is increasing. Korea's trade surplus with the U.S. was $9.5 billion in 1987, $7.4 billion in 1986, $4.3 billion in 1985, and $3.6 billion in 1984.

Because of her high dependence on trade, Korea is vulnerable to pressure from her major trading partners. The trade imbalance between South Korea and the United States has become an important source of friction between the two countries. The protectionist sentiment in the United States grows stronger against Korean products. The U.S. is also pressuring Korea to ease her import barriers and drive up the value of the Korean currency (won). Paradoxically, American pressures to open markets come at the very time of Korea's transition to a more open, freer political system, and fuel anti-American movements inside Korea.

Korean society is currently undergoing a rapid transition as its authoritarian political system is replaced by a more democratic government. Korea's industrialization has squeezed a century of the advanced countries' experience into one generation. As a result, its political and social sectors have not been able to keep pace with industrialization. In addition, there are big gaps between the very rich and the very poor, between the powerful and the powerless, between urban centers and rural peripheries, and between big conglomerates and small and medium firms. In order to build a political and social system compatible with the pace of industrialization, Korea needs massive reforms. The reforms require that politicians, conglomerates and others rethink their positions. As Korean society undergoes a political transformation, its economy is also in the process of change and readjustment. The government has announced its commitment to liberalization and structural transformation, but concrete policy measures have not yet been spelled out and one can only speculate on the outcome.

However, it is reasonably clear that the domestic political setting for economic policy-making is changing drastically in South Korea. The government, which used to be the all-powerful engine of the "Korea, Inc.", can no longer monopolize economic policy-making processes. Nor can it contain them behind closed doors as it did in the past. The process of democratization has made it more difficult for government to intervene arbitrarily in the economic activities of private enterprises. At the same time, economic policies are now subjected to the overhaul of the opposition-dominated legislature, which has become much more powerful under the new constitution. Business decisions have to be made in more open and complex environments. Changes in political settings will have significant impacts on the nation's economy and on the way economic policies are formulated.

This paper aims at probing the impact of South Korea's political transition on economic policy formation. In the following sections, I will first outline a contour of political changes in order to set the political parameters of economic policy-making and map the possible direction of change. I will then delineate a set of forces and constraints operating in the process of economic policy deliberations. This will hopefully lead to a better understanding of the various factors and forces which pull the economy toward change, continuity and stabilization. As political and economic transitions are on-going processes, and the issues and problems under investigation overwhelmingly complex, the analysis conducted herein is sketchy, and conclusions derived are tentative. By the same reason, the methodology employed is exploratory, and the data selective, less than systematic, and incomplete.

## 2. Political Transition and Reform Agenda
### A. *Recent Political Experience*

Many westerners say that political miracles have happened in Korea since 1987. In June of that year, a massive political uprising forced then President Chun Doo-Hwan to kneel down; and in a June 29th declaration, Mr. Roh Tae-Woo, Chun's hand-picked heir, made a political turnabout, and laid the groundwork for a democratic transition. (Ahn, 1988) A direct presidential election was subsequently held on December 16, 1987 with all the important political prisoners, including Mr. Kim Dae-Jung, set free to run in the election. In the heated election, Roh Tae-Woo of the ruling Democratic Justice Party (DJP) won the presidency with 37.9 percent of the vote. Kim Young-Sam and Kim Dae-Jung, having failed to present a unified opposition candidacy, lost their bids, with a combined total support of 55.9 percent.

Although the election was followed by a series of charges by the two opposition candidates that the result was a fraud, it was hailed by a majority of the people as a landmark political event that led to a peaceful transfer of power for the first time in the nation's 40 years of political history. Another surprising result came in the general election for the National Assembly which was held on April 26, 1988 under the provision of the new constitution made effective in 1987. The seat distribution favoured the three combined opposition parties, and the ruling party failed to get a majority in the legislature which had been made much more powerful by the new constitution. This situation was feared by many as one which would lead the country to a political deadlock. But somehow the system has managed to maintain precarious political equilibrium and continues to adapt to a changing situation.

The two elections and subsequent transition process revealed interesting features of the political transformation from an authoritarian system to a more liberalized, democratic form of government. First of all, the electoral process provided a social catharsis in the sense that oppressed social demands and political claims were overtly expressed and began to be articulated through the institutionalized political process. Before, they tended to be oppressed, hidden underground, and often accompanied more radical and violent movements. The process also revealed the fragile nature of the democratization process in a society which has a long tradition of authoritarianism and has experienced an extended period of military-bureaucratic domination. This fragility was evident in a number of ways: the political opposition organizations were weak and fragmented; the government was tempted to use the age-old tactics of illegal vote-buying and manipulation; the elite performance and responsibility were far behind the general expectation of the people; the grassroots bases for political mobilization and participation were very weak; and the opposition parties and their leaders failed to present themselves as a viable political alternative to the military-dominated ruling party. Finally, the transition process precipitated a significant change in the rule, method, and mode of decision-making. Of particular importance is a notable change in the settings of economic decision-making, brought about mainly by the extensive politicization of economic issues during and after the elections.

## B. Emergence of Class Politics: Critical Reform Agenda

The agenda for political reforms in South Korea is complex and involves many difficult issues. On the top of the list lies the problem of military disengagement from politics. The country also needs to overcome legacies of past political tradition and cultural mores such as the authoritarian political behaviour of elites; excessive egalitarianism on the part of citizens; centralization and bureaucratization of the government; and regional political animosity stemming, partly, but to an important extent, from imbalanced development projects. I will not, however, deal with all of these. Instead, I will focus on one issue which appears to be of critical importance to understanding the changed settings of economic policy-making in South Korea; namely, the emergence of class politics and ideological cleavages to the forefront of the political battleground. It seems that the reform agenda along this dimension is intricately linked with the solutions of such problems as politicization of labour relations and urban problems, student radicalism, unification movements and anti-American movements.

In a normal situation, elements of class conflict and ideological cleavages between the right and the left, which are usually brought about and intensified during the industrialization process, would have been solved through the bargaining or the political fighting which occurs within the context of the domestic political arena. In South Korean politics, however, the ideological battlefields for class conflict were fundamentally removed from the domestic political domain upon the onset of its modern nation-building and industrialization period. When the country was divided into the South and the North, class politics was completely prohibited in each of the domestic political domains. In the South, the leftist political movements were prohibited by the constitution, while the rightists were completely purged in the North. Accordingly, the room for any class politics has remained closed until very recently.

The consequence has been that ever since the establishment of the liberal regime in the late 1940s, competition between different political parties in South Korea over who controls the state apparatus has been among conservative factions. This pattern has remained basically unchanged until now. Therefore, there are very little ideological and policy differences between the various political parties, be they in power or opposition. What does this imply?

The process of industrialization and the attendant growth in the size of the labour movement on the one hand, and a deepening societal diversity on the other hand, have made it very difficult for the political system to steer society without resorting to a coercive, authoritarian method of political control. Phenomena of "class-based" political movements began to surface in Korean politics during the later period of the Park Chung-Hee regime. In the beginning, they took a form of loosely coordinated opposition movements aiming primarily at "political democracy". Later on, facing harsher and increasingly repressive measures, they turned into more organized, ideologically oriented, and intensely politicized movements stressing economic causes under the banner of "social justice" or "economic democracy". Park Chung-Hee suppressed them by strict legal and political means. Still, the movements became more vigorous at the turn of the 1980s, with some going underground. The fifth republic under Chun Doo-Hwan similarly resorted to coercive repressions, but with less effectiveness. However, the recent transition period and democratization process has made it much easier for the opposition groups and leftist organizations to have their voices heard, and they actually made a significant impact on the voting patterns of the last two elections. However, the formal institutional structures and traditional party organizations are not yet ready to incorporate them into the established rules of the political game.

The result is that there has been an upsurge of ideological debates and right versus left conflicts. These cleavages are manifested in student radicalism and violent labour movements. One of the gravest challenges of transition politics is how to get these movements under control, while carrying on the speedy democratization programs. The regime in charge is called upon to devise a reasonable solution for this difficult task. What is needed is to encourage the growth of the interest groups and political organizations able to absorb class politics within an institutionalized framework. At the same time, the style and operational codes of the decision-making process needs to be radically modified to adapt to changed political situations. Can the current political system afford to allow a class-based party system that would include, for example, the introduction of a socialist party? Or, should the existing party system be reformed so as to absorb the shock? Can the Roh government initiate any reform program along this direction? These are important questions, and bear critically important implications for the prospect of Korea's political economy in the 1990s.

## C. Reform in a State-dominated Society

There are certain limits to reform efforts in a country like Korea which has a long history of a strong state, dominating relatively weak, underdeveloped, civilian, social organizations. (Ahn, 1988) It is important to point this out, since a strategy of reform must take account of historical considerations. A regime can be replaced overnight, but the path to liberalization and democratization can take multiple routes. It also involves a much more complex process, which is often full of weak points and fragile steps. Radical social reformers pre-suppose that genuine reform is possible only when the political monopoly of the ruling class (or its party) is dismantled. However, the dominance of ruling class is not likely to be overthrown overnight in countries where the ruling coalition has been entrenched for a long period of time, taking advantage of weak societal forces and fragmented opposition. Transition politics in South Korea provides ample evidence to support this thesis.

As we see in the cases of Eastern Bloc and many Third World countries which have been relatively successful in a slow and gradual process of liberalization, reforms with only moderate objectives are likely to succeed. Success is usually made possible largely because these reform measures permit the continued privilege of the ruling coalition while implementing a gradual process of concessions and withdrawals. Therefore, at least in the initial period of democratization, the reform movement is likely to draw success and gain momentum when it seeks only limited objectives. At the same time, some of the reform programs can be successful only when they are effectively steered by the existing power blocs. Along this line, what is currently happening in South Korea is worthy of more careful examination.

Transition politics in South Korea have confronted many trials and errors along the way. Newly-created opportunities for self-expression and participation on the part of opposition politicians, student groups, the labour movement, and other radicals have often bred social uncertainty or required an extended period of adjustment. New groups admitted into the political system and opposition forces often intensified direct confrontation with government, testing the limits of its tolerance. When this occurred, conservatives instinctly called for reaction. Pressures coming from all sides often threatened the government's capacity to maintain equilibrium. President Roh's government has been pushed hard in two opposite directions. One pressure comes from the opposition parties and populist groups. They argue that the incumbent government is the same old group responsible for the bad things that happened in the past, and claim that the regime is incapable of keeping its promise to the people. At the same time the ruling party, and especially president Roh, are pressured by a strong conservative faction within and outside of government. They maintain that the democratization programs are out of control and are creating chaos and disorder. They charge that the government is indecisive and ineffective in dealing with the opposition "revolutionary" forces. The view is also shared by some in the military and the business sectors. President Roh has not yet proven his competence and established a style of leadership strong enough to handle these conflicting demands. Unless he meets this challenge effectively, the country may face another period of setback.

The above analysis suggests, first of all, that democratic transition is a fragile process and requires careful nurturing and support both within the confines of its borders and abroad. It also suggests that in transitions, there are elements of both change and continuity. Therefore, the transition process involves prospects for hope as well as uncertainty. In general, however, it is the contention of this paper that optimism and continuity will bear out the passage of time in South Korea. An examination of its economic prospect will lead us to feel more competent about a cautious optimism and a prediction that changes in economic policies, though they will be significant in many respects, will be brought about without a fundamental shift of South Korea's past policies and commitments, domestically and internationally.

## 3. Prospect of Economic Reforms: Continuity and Change

### A. Pull Factors for Maintaining Continuity

It is widely known that Korea's rapid industrial growth is largely indebted to a strong state, especially to its capability to channel public and private resources effectively to productive, dynamic, industrial sectors. However, the thesis of the strong state needs to be examined with respect to two restraining factors. First, the great importance of the international market for a small, open economy like South Korea sets a limit on the state's strength and its ability to intervene in market principles. Insomuch as international market constraints are largely beyond the control of the state, its freedom of "autonomous" actions in economic policies tends to be circumscribed by market principles. This in turn generates forces for continuity in economic policy and, at the same time, pushes the economy toward further liberalization and structural readjustment in keeping with international market requirements.

The second caveat of the strong state thesis lies in its inability to explain the dynamic changes in the conditions of the state interventions. (Shafer, 1988) The mode of state intervention and its effectiveness are not invariable throughout the development process. Rather, as the economy grows in size and become more complex in structure, the relationship and autonomy of the state vis-à-vis various groups and elements of a nation's political economy also change. The state's strength is also relative to different sectors, issues, and tasks involved in the making of economic policies. The net effect of the changing social and political dynamics attendent upon the deepening industrialization process will produce factors that move a society towards economic liberalization and political democratization.

*(1) Market Dependency*
The basic structure of the Korean economy is characterized by an export-oriented industrialization (EOI). Economic policies in EOI countries are determined primarily by the structure of the world market and the conditions of foreign demand. In these countries, market principles dominate the process of both domestic and foreign economic policies. In general, Korea's economic policy

follows the same pattern, albeit with different priorities and changes of nuance over time in its development sequencing and changes in the characteristics of regimes. This implies that other things being equal, forces to maintain continuity of past policies and practices will prevail over the forces pushing the economy to a discontinuity. Let us examine this hypothesis more closely.

The importance of foreign markets for EOI countries has several implications for their economic policies. First, foreign economic policies of EOI countries tend to be primarily determined by the requirements of international market forces. Second, so long as the market competition model serves as a basic framework for EOI countries' foreign economic policies, the major actors of the national economy—the state, capitalists, labour unions, etc.—are restrained in their ability to intervene arbitrarily or act against the operation of market principles. Third, the political risks, especially in the international arena, of government economic intervention are far more serious for a country depending on an EOI strategy. Because of this, the state is more cautious and restrained in manipulating economic policies for the sake of political gains. This will in turn allow policy-making in economic sectors to be more independent and autonomous of interventions originating from the political sphere.

What does the above discussion tell about Korean settings? It tells first of all that the political control of the economy is less tenable in Korea than in the case of Latin American countries where they have larger domestic markets or less dependence on world markets. It further suggests that the mode and manner of the state's economic intervention in Korean settings would be concerned more with market-protection than market-restraint. Furthermore, as political instability or labour turmoil has a direct bearing on its international standing, the state should pay more attention to such things as political legitimacy, stability in economic policies and industrial peace than those countries which are less dependent on the world market. By the same reason, economic policy-making needs to be more inclusive and tuned more to workers' docility than to the advantages of low-wages. Finally, changes in domestic political and economic conditions are likely to be managed so as not to disturb Korea's standing in international politics and world markets. For these reasons, we can see that the settings of economic policy-making in Korea are different from many Third World or Newly Industrializing Economies.

This is not to deny that economic policy is also a function of state and regime characteristics. The structure and institutional characteristics of the state (and political regime) have a significant effect on economic policies, a point to which we will turn next. In fact, the combination of authoritarian regime and EOI presents an interesting question about the nature of economic policy-making. Authoritarian control may or may not undermine market principles. So long as market principles help the political stability of the authoritarian regime, the authoritarian leadership will respect and abide by the principle of non-intervention. If it does not, the regime will manipulate the market to its favour, or attempt to suppress market principles. However, regimes in EOI countries will still be restrained in their ability to control the economy. Their freedom of action will be limited to a range of activities permissible and set by market principles. To the extent that an authoritarian leader defies the limits and constraints imposed by market principles, he will face the risk of his leadership being increasingly undermined.

In short, market dependency plays the role of facilitating economic liberalization and privatization. It also keeps economic policies of the EOI countries from abruptly disassociating with market-based past policies and practices, and thus helps to maintain basic continuity in economic policies. This is particularly so for foreign economic policies. To the extent that the Korean economy is led by export-oriented industrialization and its continuous growth remains dependent upon the world market, its foreign economic policy will conform to the requirements of market principles. Similarly, the process of change and reform will follow a mode of readjustment towards liberalization, privatization, deregulation and internationalization of its economy.

*(2) Strong State*
A country's economic development depends on a number of factors, such as favourable world economic conditions, effective political leadership and economic policies, and optimal industrial relations and financial capability. However, the nature of economic growth in many developing countries is crucially dependent upon the role of the state. The state determines which industry is to take the

lead and it oversees labour relations and the allocation of resources. The state is also capable of mobilizing investment funds (through forced savings such as taxes and inflation, foreign borrowing and financial intermediation) and channelling them selectively to certain development projects. It can create public enterprises and manipulate provisions of loans and incentives to the private sector. Indeed, theories of economic development in the Third World—state capitalism, dependent development and bureaucratic authoritarianism, as well as neo-classical theories— emphasize that the state is actively involved in economic policy-making. A dominant role of the state has often been vindicated as necessary for rapid economic growth in the Third World.

Likewise, many studies attributed the success of the South Korean economy to the "strong state" and "proper policies". (White & Wade, 1984; Ahn, 1988; Shafer, 1988) The following account is representative of this argument. Park Chung-Hee inherited a state with strong administrative and coercive machinery introduced by Japanese imperial rule and later reinforced during Syung-Man Rhee's reign and in the aftermath of the Korean War. When Park took over the government, the South Korean state was virtually "insulated" from old, traditional business organizations and societal pressures. He also benefited from his military background which allowed him to dissociate himself from former civilian politicians and businessmen tied to the "corrupt" Rhee regime. Park Chung-Hee was a strong, authoritarian leader who had a strong commitment to "save the nation from the age-old poverty and subjugation." As the business sector and workers were too weak to organize for collective actions vis-à-vis the public sector, it was relatively easy for Park to shape them to his own will and build a "developmental state" which was "autonomous" and "strong" enough to command development programs guided by the principles and priorities dictated by the state.

Empowered by the strong state, Park established strong, insulated policy-making institutions and filled them with a competent and professionally skilled cadre of bureaucrats. They were in turn given the task of formulating policies of industrialization and managing the implementation of the economic development programs. The Economic Planning Board (EPB) is a prototype of such institutions. Created in 1961, and the position of its head elevated to the rank of Deputy Prime Minister in December 1963, the EPB has been the "heart" of Korea's economic policy-making process, the "central coordination body" of the nation's economic planning, the "top economic advisor" to the president, and the "principal spokesman" on the economic policy of the developmental state. (Choi, 1987) In alliance with the economic secretariat of the president's executive office (Blue House), or often by the political mandate of the president himself, the EPB's ever-growing influence frequently pre-empted the jurisdiction of other economic agencies such as the Ministries of Finance, Commerce and Industry, Agriculture and Fisheries. The EPB's influence in economic policy-making grew stronger during the late 1970s and reached its peak in the early to mid-1980s. The recent transition of government seems to present a challenge to the central, strategic role of the EPB. However, the economic policies of the current regime maintain a basic continuity with the state-centric model. Therefore, the pre-eminent role of the state and central importance of the economic bureaucracy, such as the EPB, are a continuing phenomenon in economic decision-making.

However, as the political settings of economic policy-making undergo changes, persistent state interventions and bureaucratic omnipotence in economic policy-making begin to pose new sources of conflicts. Various forces and emerging factors will stimulate change and reform, and eventually push the economy towards speedier privatization, deregulation, liberalization, and decentralization of the policy-making process. Two structural variables deserve closer examinations in this connection. One is the logic of deepening industrialization, and the other the process of further democratization.

## B. Push Factors for Change and Reform
### (1) Deepening Industrialization

The success of state-led economic growth in Third World countries often plants the seeds of its own source of undoing. The theory on the dynamics of deepening industrialization argues that as industrialization proceeds, the power and influence of the private sector organizations—particularly industrial and financial—will also grow relative to the state and public organizations. This will in turn, limit the capability of the state to exercise its power and influence vis-à-vis private sectors and erode its relative freedom of intervention and its effectiveness.

As a country's industrial structure becomes more mature and complex, the political-economic environments surrounding the state, industries, financial institutions, and the labour movement usually undergo significant changes. The structural conditions of the society and balance of power among these groups and their organizations will also shift accordingly. Therefore, the method and manner in which the state intervenes in the economy needs to change as industrialization proceeds. Otherwise, it is likely that the degree and intensity of the conflicts between different social groups and classes will become more serious. This may in turn erode the legitimacy of the state organizations and bring about more organized opposition to the existing regime and its ruling coalition. Instances of political instability, authoritarianism, and regime changes in many newly-industrializing Third World countries have been explained in terms of this logic.

The recent Korean experience has a strong resemblance to this generalization. As the size and complexity of the economy increased, the private sector and social groups became increasingly vocal about the negative aspects of the state-centric, economic development policy. The social criticisms and political attacks against state interventions in the economy also rose sharply. For example, industrialists protest more frequently and vehemently against the persistent intervention and increasing role of the state in the economy. State-owned or state-dominated public enterprises are increasingly under pressure to be privatized. In addition, bankers oppose the regulation of interest rates and credit allocations by the state. At the same time, as the number of labourers increased, the spread of labour organizations grew and began to challenge the industrial and political peace of society. Over the years, these unions increased their collective power vis-à-vis enterprises and the state. The farmers, who were perceived as the most politically acquiescent group, have also begun to be outspoken in their demand for a higher share of the profits of economic growth.

All these changes eventually joined forces to press the authoritarian government to adopt a policy of relinquishing many of its important levers of interventions in the allocation of economic resources. Government is also committed to implementing measures aiming towards speedier economic liberalization, structural readjustments and, eventually, political democratization. Liberalization and democratization programs undertaken so far have brought the social and economic demands of various groups and classes directly on to the central stage of the political process. The government is no longer able to impose price or wage controls, and finds itself unable to take decisive actions to the extent that was possible under the previous authoritarian rule. With these considerations in mind, we may confidently predict that the logic of deepening industrialization is positively correlated with mounting pressure within society to change and reform its economic and social structure toward more open, liberal, and democratic structures. As political openings continue and democratization proceeds in South Korea, the autonomy and freedom of its government to intervene in societal sectors will also decrease. In short, the traditional, state-dominated mode of policy-making in economic affairs is becoming less tenable.

*(2) Impacts of Political Democratization*
The recent transition toward democracy have brought about major changes in the political environments in which economic policy is now made in Korea. The consequences of the changes will have direct impacts on the manner and mode of its economic policy-making in the following three ways.

First, the transition brought about the entry of broad "democratic" opposition groups into the political arena. Further openings and liberalization of political activities provide these groups with new channels to express their demands and grievances. As a result, the process of economic decision-making is subject to more open public criticism. It has also been impelled to become more inclusionary, that is, allowing more groups to participate in the process and share in the benefits.

Second, the logic of liberalization and democratization has made the government's use of coercive measures increasingly difficult and politically costly. The opposition parties, operating in a more open political context, have been empowered to exploit, more easily than before, the opportunity to challenge the orthodoxy of government's economic policy. Economic issues therefore tend to be politicized more easily and more frequently. As a result, the previous domination by the government and ruling coalition of economic policy-making has been significantly eroded.

Third, open and democratic environments often intensify conflicts between the political elite and the technocrats. Facing the electoral challenge, the political elite, especially the ruling party, began to be more concerned about rebuilding political support through changes and manipulations of economic policy instruments. This in turn brings about a decline in the independence of the economic bureaucracy and intensifies competition and conflict among the major economic policy-making institutions.

These changes are significant and imply a major source of conflict and discontinuity between the past practices and the future requirements of economic policy-making in South Korea. Previously, the economic policy-making process has been dominated by the executive branch which has been shielded from political criticism and legislative interference. With new environments, however, decision-making has now to navigate a more complex and politicized process. The opposition parties, outnumbering the ruling DJP in seats, currently control nine of the 16 standing committees of the legislature. They include important economic committees such as industry, energy, transportation, construction and economy-science. President Roh and his economic ministers can no longer "rule by fiat". They have to deal with an opposition-dominated legislature and confront client groups who are much more volatile.

When the senior bureaucrats in the EPB and Ministry of Finance "lobby for support from a man who was tortured and jailed only a few years ago" (Opposition leader, Kim Dae-Jung),[1] the changed political environments has already begun to "put its imprint on economic policy-making". Inter-bureaucracy competitions and jurisdictional conflicts among different economic agencies are also likely to arise more frequently. For example, when the EPB recently pushed a policy to accelerate liberalization programs in the finance and import sectors, the Ministries of Finance, Commerce and Industry, and Agriculture and Fisheries all rose to openly resist the policy. It has also been reported recently that policy differences are becoming wider and interest conflicts deeper between the EPB and various economic ministries such as the Construction Ministry (over housing policy), the Transportation Ministry (on raising taxi fares), the Agriculture and Fisheries Ministry (on opening up the agricultural commodity markets), the Finance Ministry (on currency control), the Energy Ministry (over lowering oil prices), and the Home Affairs Ministry (on land tax policy).[2]

Viewed as a whole, however, economic policy in South Korea is still directed largely by the state and this trend is not likely to reverse its course in the near future. The state-centric, macro-economic model, which is geared to international market principles, is still widely viewed as right. What is needed in Korea is a gradual readjustment of the economic structure and reordering of the policy objectives to fit the economy to the requirements of international demands on the one hand, and the changing domestic political settings on the other hand. The trend to maintain continuity will go on because of the effects of past policies and habits of the major economic actors. For example, businessmen keep coming to government for a helping hand to secure their comparative edge in international markets. Fearing that the emerging labour movements will undermine their interests and comparative advantage, they call for tougher measures to curb strikes. Big corporations will beg government to provide financial assistance to save them from bankruptcy. Workers look to government agencies for protective interventions. Bankers expect the state to come up with guidance before they decide what to do.

The preceding arguments can be put into the following hypotheses:

I. The higher the degrees of international market dependency and openness of a country's economy, the more difficult it is for the state to intervene effectively in the economic policies and dissociate itself from policies and practices of the past which were built upon the requirements of market principles. At the same time, the higher the market dependency, the more there will be pressures, coming from both within and outside of the national economy, on the state to adopt policies to move towards economic liberalization, privatization, and structural adjustments.

II. Other things being equal, as the process of deepening industrialization and economic liberalization proceeds, there will be demands and pressures for political democratization. To take this one step further, as liberalization and democratization proceeds, there will be less room for authoritarian government to monopolize economic policy-making or take repressive measures against market principles.

## 4. Case Studies: Actors and Processes at Work
### A. *The Wage Policy*

In South Korea, like many other countries, two parties have been primarily involved in making decisions and policies regarding wages: the state machinery and trade unions. The EPB is the peak state organization which lays down basic guidelines on wage levels and annual increase rates for workers in both the public and private sectors. The wage guidelines of the EPB are set in line with annual growth targets and the inflation rate. Another state organization involved in wage policy is the Ministry of Labour, which, until 1981, was called the Bureau of Labour Affairs, but was subsequently elevated to a ministerial level, reflecting the growing importance of labour issues in South Korea. The primary role of the Labour Ministry is the supervision of the state wage guidelines set by EPB.

On the union's side, the Federation of Korean Trade Unions (FKTU) sets its own guidelines and targets of annual wage growth for local unions to follow before wage negotiations begin every year. The FKTU's annual targets have been considerably higher than the EPB's guidelines. The FKTU's role has been variable over time as issues change and political environments evolve. But, in general, its role and influence has been restricted and permitted only in a prescribed range of activities imposed by legal or political considerations.

According to labour law, which is still in effect though being criticized as antiquated and repressive to workers, trade unions in South Korea are permitted to organize at the enterprise level only. Only one union is allowed in each company. Each company union must affiliate with one of 20 national union federations. In the past, these federations have been regarded as government-controlled organizations. The wage negotiation is primarily conducted between employers and enterprise union leaders at each enterprise level. Because of enterprise unionism, the wage negotiation process in South Korea has been characterized as "diffused and decentralized" so that workers and unions remain relatively ineffective in dealing with owners and state organizations.

Two primary objectives have been taken into consideration in determining wage policy. On the one hand, the state tends to keep wage increases lower than productivity increases so as not to erode the country's competitive edge. On the other hand, the state is responsible to protect the workers in the sense of securing a reasonable rate of wage growth. The wage is taken in this case as an important source of material rewards to enhance the legitimacy of the state and sustain the political support necessary to maintain social stability. The state's guideline for wage determination is made in such a way as to balance the two conflicting policy objectives—that is, the wage-restraint function and the wage-protection function. If the state fails, for example, in the wage-restraint function, the economy will suffer from a loss of its competitive advantage in the world market. If it falters in the wage-protection function, the system cannot secure continuous sources for productivity and/or the state will face an upsurge of worker's discontent.

In the past, South Korea's wage policy had been concentrated on the wage-restraint function for fear of erosion of its competitiveness. Considering its high dependence on the world market and relatively unfavourable technology endowment, it was quite logical to preserve its comparative advantage by maintaining low labour costs. However, it needs to be underscored that the government's ability to restrain wage increases in the Korean setting, with its export-oriented industrialization, has not been as strong or effective as in import substitution, industrialization countries.[3] It is true that the wage policy in Korea has been repressive and unfavourable to workers. But, when the political risk of wage restraint became so severe that it endangered the basic stability of the economy or the regime in power, the state often resorted to measures to suppress the corporate interest in favour of wage raises for workers. Once the industrial system failed to contain labour disputes within a permissible range, or when the conflicts escalated into large-scale political uprisings or became susceptible to "class conflict", the state economic policy paid more attention to industrial peace than to the advantages of low wages.

As the process of liberalization and democratization accelerates, the state is being pressed increasingly to reverse the priority or, at least, change past practices in setting wages and other policy matters relating to labour relations. The state machinery in South Korea is still a dominant authority and the final arbitrator of labour disputes. But, as the state's supremacy over society is challenged during the transition toward democratization, the political costs entailed in it's wage-restraint function are also increasing. A repressive wage policy primarily based on the principle of comparative advantages has become less tenable. Let us look at some facts bearing on this point.

During the summer of 1987 Korea underwent an unprecedented period of labour unrest. Over three thousands factories were hard hit by violent labour uprisings. The strike count was 3,800 in 1987. In previous years there had seldom been more than 100 strikes a year. Workers, of course, demanded more pay. But, the primary target of the labour movement was directed to the state's intervention against it. Workers demanded that the state cease its interventions in wage settlements and that "free unions" be allowed. Workers gained a fair degree of state concession and the net wage increase amounted to 13.5 percent which, when added to the earlier 9.1 percent wage increase in the Spring Wage Raise, amounted to a total wage raise for 1987 of 22.6 percent. A second wave of labour unrest almost hit the country in spring 1988, being touched off by strikes in the auto-making and shipbuilding industries. But the unrest did not spread widely and was confined to industries owned by the big conglomerates.[4]

Part of the reason for the relative industrial peace in 1988 might have been that the new government has dealt more effectively with the workers. The more important reason may be, however, that workers and their unions got what they wanted without resorting to the same tactics that brought success in previous years. For example, the production workers easily won wage increases as high as 15 percent in an earlier round of wage negotiation in 1988. The wage growth in Korea during 1987-88, which according to the *New York Times* (February 19, 1989) was highest in the world, outstripped productivity gains, reversing the trend of 1980-86. This and other indicators show that the strength of workers and unions in South Korea is gaining rapidly relative to that of employers and the state.

The trend of wage increase outstripping productivity gains invited growing concerns and, in some sense, new thinking. A growing number of economists have begun worrying about the competitiveness of Korea's exports. The view is taken as valid by many, especially since the Korean won has appreciated against the dollar even more quickly than the currencies of Taiwan and Japan. Likewise, businessmen complain that the combination of a wage up-lifting, currency appreciation and lost production is reducing their competitiveness and slowing down the economy. They call on the government to step in and curb strikes. On the other hand, people in pro-labour camps countered that business must learn to cope with justifiable requests of the workers for better wages and working conditions. They say that Korea's industrial products are still competitive, pointing out that the country ranks among the world's highest in average working hours per week[5] and that its labour costs remain relatively low.[6]

With democratization programs in train, trade unionism in South Korea is expected to grow ever stronger. From 1987 to 1988 the union movement membership increased from 1.1 million to over 1.5 million and the number of local branch unions jumped from 2,658 to 4,729 during the same period. (*Far Eastern Economic Review*, September 22, 1988: p.30) Now that unions have been organized in most of the big companies, including Hyundai, Samsung, and IBM which used to maintain vigorous anti-union policies, unionism is likely to be far more vocal than before. In addition, many of the new union leaders (and their followers) are younger, more outspoken, tougher, radical, and often more ideologically oriented. The leadership of the government-sanctioned FKTU has recently found it difficult to maintain control of the union movement because of the challenge of younger members to the incumbent leaders. The younger trade unionists criticize the FKTU for its cooperation with the administration of former president Chun Doo-Hwan. Radicals go further and ask for rights to elect their union leaders through direct elections, opposing the current delegate system which favours incumbents. The new generation also demands unions to be more democratic and independent of the officially sanctioned union structure.

Compared to other industrialized countries, South Korean unions are still small in organization and relatively weak in resources. Out of a 9.5 million workforce, only 16 percent is unionized. But the political clout of the unions has been significantly enhanced in recent years. They endorsed or strongly supported at least 10 candidates for parliamentary seats in the legislative election held in April 1988, despite the fact that unions are not legally allowed to engage in political activities. Nevertheless, seven out of the ten won seats, without legal repercussions. With this political clout, the revision of the "antiquated" current labour law has again been placed on the priority list for legislative action. The opposition parties strongly support the revision of the labour law, which is now being drafted. It is widely expected that the new labour law will scrap the present ban on political activity by trade unions. Union-backed, pro-labour camps also pushed successfully to have the government adopt a minimum wage policy in 1987.[7] Hence, the trend is that unions in Korea will engage in a wider scope of political activities.

What is the future of labour politics in South Korea? Given that government is certainly not capable of enforcing the wage-restraint policy to the extent that was possible under repressive authoritarian rule, will the wage of workers go up continuously so as to outstrip the productivity gains and Korea's competitive edge in the world market? Will labour conflict threaten the basic stability of the Korean economy? Any attempt to provide exact answers to these questions involves a degree of uncertainty, and may suffer from risks entailed in the occurrence of unexpected events. However, our analysis offers the following tentative assessment. First of all, it appears that as the democratization process takes root, labour's share of political influence and economic benefits will increase. Second, the ceiling of the workers' share and the state's wage policy will depend on the limits and requirements imposed by the world market principles and the state's ability to maintain its supremacy over the society— particularly its influence and effectiveness in maintaining the balance of power among major groups involved in the process of setting wages.

## B. *Financial Liberalization Policy*

South Korea has pledged to join the international movement toward financial liberalization. However, progress to date has been very limited. Several years ago the government announced a four-stage program of international liberalization of the financial markets, but it has been unable to keep its promises, not because it is unwilling, but because financial liberalization has to first develop domestically. Korean financial markets have been very carefully regulated by the government. The Ministry of Finance (MOF) is the key institution which oversees the regulatory and the control structure of the financial sector.

Since the beginning of its industrialization drive in the mid-1960s, Korean government has exercised tight control over the financial system. A high rate of investment spending has been key to rapid economic growth. Investment requires inducing domestic savings and procurement of foreign borrowings. In order to secure a high rate of investment, government exercised a broad financial regulatory power and control over the functions of the financial intermediaries, manipulating the investment criteria in a number of ways.

When the country embarked on its industrialization plans, the profit and security criteria of financial institutions did not often go hand-in-hand with the development priority set by the state. Therefore, the possibility of resource mobilization through market mechanism was considered insufficient for the projected industrial growth. It was in this regard that the state's intervention in the financial market was necessary and desirable. One mechanism of control was credit allocation policy. The state used various regulatory methods to influence credit allocations of the financial institutions. The use of taxation power, which allowed the state to manipulate the sources and allocation of its revenues, was another common tool for influencing the allocation of financial resources. Tax reforms in Korea have often been introduced in order to increase government revenues, which in turn allowed it to provide fiscal incentives to certain industrial investments.

The state also tightly regulated the monetary policy, particularly the supply and flow of money. Since the appropriate supply of money is a necessary condition for industrial development, control over it was a critically important tool of the developmental state. Finally, the state had various means and provisions of subsidies at its command. They included tax breaks, low interest funds, low-priced inputs, and other financial incentives or preferential treatments. The state used these subsidies to induce private investment in selected industries so that the growth rate of those industries would be higher.

The state's capacity to regulate the financial market is a function of its relative strength vis-à-vis the society in general and the institutional (collective) development of financial intermediaries, in particular. The state's strength varies as a function of time and changes in the capacities of its constituent actors— industries, financial intermediaries and trade unions. Therefore, we expect that as the financial institutions grow in size and complexity and/or their ability to mobilize societal support for autonomous decisions rises, the state's capacity to regulate the financial market will also become less tenable. The changing pace of the state-society relationship in South Korea provides an ideal case to examine this phenomenon in a dynamic perspective.

In earlier days, financial institutions in Korea were small in number and weak in structure. But since 1980, the financial sector has grown rapidly, improved its efficiency, and deepened the diversification of its products and services (Park, 1987: 176-189). The impressive growth and maturity of the financial sector has been possible in spite of the continuation of repressive financial policies.[8] However, as the economy grows bigger and more complex, government's financial interference has increasingly restricted the country's ability to move effectively toward an outward-looking, export-led development strategy. Also, there were other economic and political complications that made Korea's economic prospect quite pessimistic and convinced the public and policy-makers that a major policy shift was essential. Paramount among them was a backfire resulting from the massive investment in heavy and chemical industries of the mid-1970s and consequential financial difficulties; soaring inflation and mounting deficit accounts coupled with poor harvests in 1978 and 1979; the first (in 1973) and second (in 1979) oil-shocks; the world recession and the erosion of Korea's export competitiveness; and the huge accumulation of foreign debt. In addition, the political situation was quite volatile in the aftermath of President Park Chung-Hee's assassination in October 1979. It was against this background that reform measures were introduced in 1980 to reorient financial policies towards stabilization, deregulation and liberalization.

For the last several years, there have been a number of policy attempts to liberalize financial markets in line with the basic objectives of "more openness, greater autonomy, and decentralization" of the economy. However, the achievement has not been substantial. A full documentation of reasons for the faltering achievement may be necessary in order to better understand the dynamics of the liberalization process. But the single most constraining factor that has hampered the liberalization process has been, as Park concluded, "the absence of a constituency strong enough to push through the liberalization programs." Park goes on to say:

> Not surprisingly, neither the business community nor the bureaucracy has been receptive to the liberalization program. Consumers, who could benefit most from the liberalization, have not been able to organize themselves. (Park, 1987: 203)

Again, the prospect is that the changes will come but the basic continuity will not be scrapped: the financial market will still be controlled largely by the state authority; government's financial policy will continue to function as a major instrument to achieve state-directed developmental goals; and any change that may be introduced will be gradual and piecemeal. This does not necessarily mean, however, that things are quiet, or no reforms are foreseeable. Conflicts of interests between government and major financial intermediaries are building. New legal measures are being politically debated between the government and the opposition.

For example, the amendment of the law governing the Bank of Korea has recently become a hot political issue. In the past, the Bank of Korea was tightly controlled by the Ministry of Finance. As a result, the bank has been used to pump money into the economy during elections and to give favours to politically-connected corporations. Such political abuse of the bank will, however, be impossible in the future. The opposition now wants to make it an independent institution. They also demand that the bank's monetary policy board be independent and filled with more outside representatives, including those from small business and farming. The Bank of Korea appealed to defer the amendment issue until next year to avoid entanglement in partisan politics.

The incident shows that the politicization of the economic policy agenda may entail complications as well as opportunity.

The opposition parties have not yet come up with a coherent economic policy program of their own. However, they have called for a major shift in the country's economic policy. They maintain that political favouritism must stop. They also demand that the central financial institution (Bank of Korea) be made independent from government. In principle, the path advocated by the opposition groups calls for reforms in the following direction:

1. A more cautious approach toward participating in the world economy;
2. Being less friendly to the penetration of certain transnational actors;
3. More orientation toward domestic rather than foreign markets; and
4. More concern with equity and social balance than growth and expansion.

The shape and prospect of Korea's economic policy will depend a lot upon how and to what extent the current economic system can incorporate those challenges.

## Notes

1. Quoted from, *Far Eastern Economic Review*, August 4, 1988, p. 44-45.
2. See report of *Chosun Ilbo* daily, February 22, 1989.
3. In this respect again, Korea is different from many of Latin American countries which have been implicated by such theories as bureaucratic-authoritarianism or dependent development.
4. According to the Ministry of Labour, the labour dispute count for 1988 was 1,873.
5. The average working hours per week among Korean industrial workers in 1988 was 54, compared to 41 in the U.S. and Japan, 40 in West Germany, and 48.1 in Taiwan.

6. For example, the average hourly wages for auto workers are $3.60, compared with about $18 for Japanese workers. (*New York Times*, September 29, 1988).

7. The exact figure of the minimum wage has not yet been set. The current proposal calls for about $1.50 an hour.

8. Cole and Park's study revealed that although Korea's financial system was "not always effectively managed," the Korean economy performed remarkably well. On this basis, they concluded that "the Korean government used the (financial) system effectively to transform Korea from a poor agricultural economy to a newly industrialized economy." D.C. Cole and Y.C. Park, *Financial Development in Korea, 1945-1978*, (Cambridge: Council on East Asian Studies, Harvard University, 1983), quoted from Park, 1987: pp. 190-191.

# References

Ahn, Chung-Si. "Korean Politics in a Period of Transition," in John W. Langford and K. Lorne Brownsey, eds. *The Changing Shape of Government in the Asia-Pacific Region*. (Halifax: The Institute for Research on Public Policy, 1988), pp. 21-42.

Choi, Byung-Sun. "The Structure of the Economic Policy-Making Institutions in Korea and the Strategic Role of the Economic Planning Board." *The Korean Journal of Policy Studies*. (Seoul: Graduate School of Public Administration, Seoul National University, 1988).

Park, Yung-Chul. "Financial Liberalization and Prospects for Resource Mobilization in Korea," in Sung-Joo Han and Robert J. Myers, eds. *Korea: the Year 2000* (Carnegie Council on Ethics and International Affairs, 1987), pp.175-205.

White, Gordon and Robert Wade, eds. *Development State in East Asia: Capitalist and Socialist*. (The Institute of Development Studies, Bulletin No. 15, 1984).

Shafer, D. Michael. "Sectors, States and Social Forces: Korea and Zambia Confront Economic Restructuring." *Comparative Politics* (forthcoming).

*Far Eastern Economic Review*. (Hong Kong, August 4, 1988).

*New York Times*. (September 29, 1988).

# China

## The Process and the Problems of the Decentralization of China's Economic Decision-Making Power: 1978-1988

*Cao Yuan-zheng*

China's economic structure has been radically reformed since the 3rd plenary session of the 11th Central Committee of December 1978. The reform was conducted on a highly centralized planned economy (the Soviet model) and its objective is to form a planned commodity economy which effectively combines markets and planning.[1] The process of reform is fundamentally marked by decentralization of economic decision-making power with priority given to the interests of economic bodies. As the power of decision-making transferred downwards, new decision-making bodies were established. Local governments, industrial and commercial enterprises, social organizations, advisory organs and individuals began to exert an increasingly greater influence on economic decision-making, which in turn led to changes in economic, political, social and cultural structures at different levels. This paper will examine the significance and problems of the decentralization of China's economic decision-making power.

## 1. Historical Causes

With its founding in 1949, the People's Republic of China, like a number of other developing countries, followed the Soviet model and established a highly centralized planned economy. The system was based on the state ownership of the means of production. Some private sector activity existed in the handicraft industries, retail business and agriculture, but it was not allowed in other sectors. In the industrial sector, the ratio of state-owned to collective enterprises was overwhelming. The economic system had two key characteristics which aligned with the state's ownership of the means of production.

(1) The planning system placed heavy reliance on spiritual stimulation as a means of encouragement. It stressed dedication, voluntary labour and the subordination of personal interests to social interests and seldom used material incentives such as wage differentials and monetary reward. As a result, individual influence on social and economic activity was minimal.

(2) In the allocation of resources, the planning system relied on the role of "the visible hand" and ignored the role of the "invisible hand". Economic coordination was conducted by governments at all levels and through the central planning process. This kind of planning focused mainly on materials and the price of production inputs and consumer goods and the control of foreign trade. At the same time the planning of finance through monetary and credit systems was limited.

The combination of these factors constituted a command economy characterized by mobilization in which instructions are imposed from above through an organized pyramidal system. On the bottom line of the pyramid are the plans of the industrial enterprises which are coordinated with the central plans through two channels: (1) the plans formulated by industrial sectors are transmitted to the specialized economic organizations directly under the sectors, which in turn transmit the plans to the individual enterprises; and (2) the provincial plans are transmitted to the prefectures and counties in the province and then to

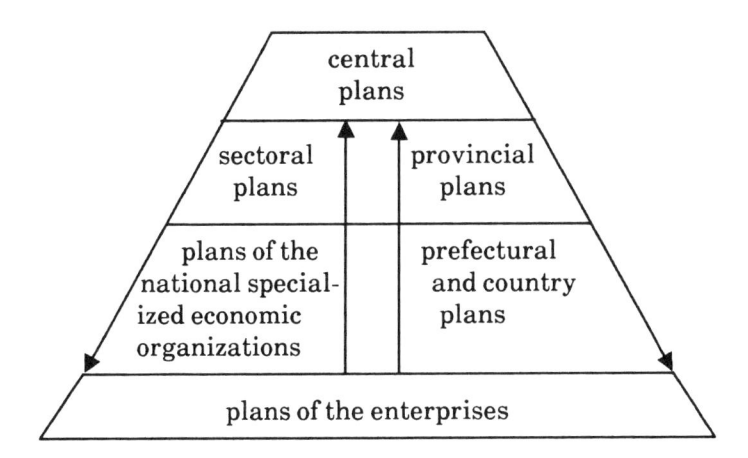

```
                    central
                    plans
         ┌──────────┬──────────┐
        sectoral    │  provincial
         plans      │    plans
      ┌────────────┼────────────┐
      plans of the │  prefectural
     national      │  and country
     specialized   │    plans
     economic      │
     organizations │
   ┌──────────────┴──────────────┐
          plans of the enterprises
```

the individual enterprises. At the same time, and through the same channels, the plans of the enterprises on the bottom line of the pyramid feed into the State Planning Commission on the top by way of these middle-level planning mechanisms. The production units or enterprises establish contact with the industrial centres or other enterprises in the region as well as establishing contact with enterprises in other economic sectors or regions. Thus, the State Planning Commission on the top of the pyramid links closely with the sectoral and regional plans and a national plan is formulated. Once formulated, this plan has a commanding nature and allows no violation. Under this system the production units and enterprises on the bottom of the planning system have no decision-making power, only the obligation to report to the higher planning bodies. Decision-making power is resident at the highest level and the middle-level has much less authority. In this type of system, the freedom of consumption and profession is strictly limited. Finance, manpower and materials are allocated according to a priority established by central planning authorities. Other social bodies, economic organizations and individuals have no decision-making power and must obey the plan.

It should be pointed out that historically such a system was in line with the internal needs of China's economic development. Pounded by the global trend of industrialization, China was compelled to take this road in the middle of the last century. Compared with the developed countries, China had a weak industrial base and very limited economic surplus. The key problem was how to best use this limited surplus for the maximum accumulation of industrial capital. The highly centralized economic planning system met the requirement. It did, however, put the consumers at a low standard of living by diverting the development resources of the country to heavy industry.

As time went by, China's economy experienced significant development and the inadequacies of the highly centralized planned economy were exposed. In a strict sense, two prerequisites were needed if the planning system was to function effectively on a permanent basis: (1) people must have boundless political enthusiasm so that they would voluntarily abandon material interests and the freedom of consumption and profession; and (2) the central planning bodies must have unlimited insight and omnipotent power so that they could direct economic activities without making the slightest mistake. However, the system is not able to create or sustain these two prerequisites by itself. The long-standing ignorance of people's interests and the limitation of their freedom of consumption and profession are bound to decrease initiative. Furthermore, the central planning bodies do not have omnipotent power and with the acceleration of the process of industrialization and the increasing socialization of economy, the possibility of making mistakes in decision-making is bound to increase.

The process of the development of China's economy has proved this point. The founding of PRC in 1949 achieved the demand for national independence and people had immense political enthusiasm. At that time, the degree of China's industrialization was rather low and its economic structure and activities were comparatively simple and easy to conduct. Therefore, the centralized planning system was effective. It coordinated with the development process well and resulted in the unprecedented success of China's five-year plan from 1953-57. After 1957, however, things began to change and the development of the economy was paralleled by a decrease in the peoples' political enthusiasm and an increase in the faults of the decision-making process.

It was against this background that the appeal for decentralization began to appear and China began to experiment with the process.

Table 1 shows that in the period from 1949-1976, the allocation of China's economic decision-making power experienced four stages of change, with power alternating between decentralization and centralization. The trend is that power decentralization tended to be of short duration and incomplete while power centralization tended to be of a longer duration. The reason for this lay in the fact that the dimension and severity of the inadequacies of the highly centralized planning system were not fully exposed and people were still of the opinion that the system itself was good and that it was only the way in which the plan was carried out that was unsatisfactory. Therefore, the key to the solution of the problem lay in the constant strengthening of the two prerequisites on which the sound functioning of the system relied.

**Table 1**
**The decentralization of power**
**1949-1976**

" + " " - " indicate the increase and decrease in the economic decision-making power.

" 0 " indicates that the power remains approximately unchanged.[2]

|  | before 1957 | 1958-1960 | 1961-1965 | 1966-1976 |
|---|---|---|---|---|
| Central bodies | + | - | + | - |
| Local bodies | - | + | - | + |
| Enterprises | - | + | - | 0 |

First, the decrease in peoples' political enthusiasm required more effective political and ideological work on the part of the party and the government (strengthening of the spiritual incentive). Second, the faults in the centralized decision-making process required a better-organized plan (perfection of the method of planning). Therefore, before 1978, the system operated in vicious circles. After 1957, on the one hand, political movements occurred repeatedly in China with the hope that they would purify peoples' consciousness and arouse their dedication. On the other hand, mandatory planning was strengthened at each link, and at the intermediate links in particular. Politically, the results of constantly strengthening the spiritual incentives led to the disastrous Cultural Revolution which deprived the country of a normal social order and plunged it into chaos. Economically, the constant enforcement of the planning system further weakened the decision-making power of the production units and enterprises, deepening their reliance on planning and changed them into accessories of the plan. Therefore, rigidity instead of flexibility, and poor results and serious waste occurred, and it was difficult to sustain economic development. By 1978, the highly centralized system of the planned economy made little contribution to the economic development of China. In fact, it had become detrimental. Further development presupposed reform.

The correct direction of the reform is to allow the economic organizations to decide, according to their own interests, what they produce, for whom they produce and how much to produce. This indicates a move towards the model of market economy. From the decision-making point of view, to push a planned economy towards a market economy, the decision-making power of the enterprises must increase. Since 1978 the system has been advancing in this direction, with the year 1984 proving to be a watershed that separates the period into two stages.

## 1. Power Decentralization Within the Original System (1978-1984)

In 1978, a fundamental problem facing China's economy was the lack of initiative for the promotion of economic development on the part of governments at all levels, production units, enterprises and individuals. The improvement of the means of stimulation and the mobilization of people's initiative became a task of highest priority. It was against this background that highlighting material interests and the redistribution of such interests, namely, structural reform in the field of distribution, became a point of departure in the reform process. In rural areas, the contract responsibility system linked with output was carried out. That is, on the condition that the state-ownership of land remained unchanged, the right to use land was contracted to the peasants. Therefore, a direct linkage was established between the income of peasants and the results of their agricultural management. In urban areas, the highly centralized financial system in which revenue at all levels was delivered to the central body, who in turn would disburse it in a unified way, was changed. Central and local revenue was separated. The practice of leaving a certain amount of revenue at the disposal of local governments appeared. This was a systematized change, based on the premise that as the local economy developed, the total finance left at its disposal should correspondingly increase so as to mobilize their economic initiative. Furthermore, a system of leaving a percentage of profits at the disposal of the state-owned enterprises, instead of being completely turned over to the government, was established. The more profit the enterprises made, the more they kept at their disposal, thus stimulating the profit-making motive. Inside the enterprises, moreover, such material incentives as disparity in wages and bonuses was employed and the total amount of wages and bonuses were linked to enterprise profits. Therefore better enterprise results would result in more wages and bonuses and would mobilize the initiative of the workers.

From the viewpoint of decision-making, the changes resulted in power decentralization from top to bottom within the original system of social and economic organizations, with the purpose of improving their functions. This purpose has been achieved. In government at all levels, the focus of work was explicitly defined as economic construction instead of political mobilization. Governments began to emphasize expertise and proposed the specialization of cadres. A vast number of intellectuals with specialized knowledge began to participate in the work of the government. This improved the structure and quality of government and policy-planning. In governments at different levels, slogans such as "Efficiency is life" and "Time is money" were proposed and accepted. Regional economic development strategies were formulated one after another, and in enterprises, factory management began to play a central role. Following this change, economic accounting was reinforced with technical personnel, represented by economists, accountants and engineers, playing a major role in the decision-making of the enterprises. This reflected a major change in managerial thinking and objectives. In the past, the production of the enterprises was simply to fulfil the state plan. According to an investigation of factory directors of 359 state-owned enterprises conducted by the Research Institute of China's Economic Structural Reform, among the 14 listed managerial objectives for the enterprises in 1984, number one was the improvement of individual benefits. Following this was the improvement of the quality of products, the development of new products, technical upgrading and the increase of profits. The fulfillment of the state plan dropped to 11th place.[3]

At the same time, with power decentralization and the general increase in the income of individuals, the rights of individuals as consumers were restored. In 1984, while conducting a reform of prices, China abandoned a rationing system for the prices of consumer goods and introduced the dual-track system, which meant that the production of goods which exceeded the state quota could be sold in private markets. According to an investigation conducted in 1985, people were very happy with the fact that they "can buy some goods though the prices are a bit higher" and they thought that this in fact improved their living standard.[4] This perception was ample reflection of people's assessment of the reform.

China's economists define this reform as one marked by the simplification of administrative control and power decentralization. It simplified and decentralized the decision-making structure and improved the quality of decisions. From 1978 to 1984, the growth rate of the national economy gained momentum and averaged over 10 percent annually.

## 3. Power Decentralization Within The Original System and the Emergence of New Economic Decision-making Bodies (1984- )

The decentralization of China's decision-making power after 1984 is marked by the co-existence of two tendencies. On the one hand, the power decentralization from top to bottom within the original system of social and economic organizations went on unabatedly. On the other hand, a large number of new economic bodies with decision-making autonomy emerged.

The so-called new economic decision-making bodies came from two sources. One is a variation of economic organizations in the original system and the other is the emergence of organizations outside the original system.

With regard to the first source, the economic organizations in the original system began to experience autonomy with the development of power decentralization. They increasingly called for a decrease in government administrative interference and began to conduct various economic activities in light of their own interests. Such changes mainly happened in state-owned enterprises. Since 1987, responding to the general demand of the enterprises, the government has followed the example of rural areas and introduced a managerial contract responsibility system in industry. Namely, government, as the owner of the enterprises, signs contracts with the state-owned enterprises, as the managers. In the contracts, government only sets the targets of profit, output value and technical upgrading. All the managerial activities are decided and conducted by the enterprises. Meanwhile it is stipulated that the people to whom the state-owned enterprises are contracted should be selected only through competitive public bidding and that it is necessary to examine their qualifications, focusing mainly on their management capabilities. Competitive public bidding can contribute to the expansion of management capabilities in the leading group of enterprises and has brought about improvements in their structure and actions. The contract system has been introduced to over 80 percent of the state-owned enterprises. In a certain sense, these economic organizations still follow the norms of the old system and haven't broken completely from the government and are therefore described as variations instead of new decision-making bodies.

New economic organizations have, however, emerged in the process of reform and are independent of the original system with separate interests. They include township and village, collective, individual, and joint enterprises.

(1) **Township and village enterprises**—These are the main elements of the new economic organizations. With the implementation of the contract responsibility system in rural China in 1978, initiative for farming was greatly mobilized and agricultural output increased by a large margin, as did peasant income. Between 1979 and 1985, the total assets of agricultural households

increased 2.68 times, for an average annual growth rate of 17.9 percent. In comparison with the increase in the currency income of peasants, activities in the agricultural economy appeared too narrow and, until the end of 1987, 75 million peasants had left agricultural for non-agricultural trades. They raised money to establish a large number of township and village enterprises. These newly emerged and spontaneous economic organizations basically act according to market principles. They neither rely on the government for the allocation of finance, raw materials, equipment and labour-force, nor are they subject to the direct interference of governmental planning. Therefore, they constitute new decision-making bodies free from the original system of social and economic organizations.

(2) **Collective enterprises**—These are enterprises financed and established by more than two people. The only difference between them and township and village enterprises is that the latter are established by peasants in rural areas or in small towns, while collectives are established by urban inhabitants in cities. They existed before 1978 as a supplement to state-owned enterprises, but after 1978 they were greatly increased, both in number and scale, and became an important force affecting economic activity.

(3) **Individual enterprises**—These are enterprises financed and established by one person only. They include the enterprises established by urban inhabitants, rural households and foreign businessmen who have 100 percent ownership of their enterprises. Since this kind of enterprise is contrary to the principle of a highly centralized economy, prior to 1978 they were regarded as capitalist and rejected. After 1978, individual enterprises gradually began to appear and grow in number. Governed in a more complete way by the law of market, they rely only on their independent management, and therefore constitute an enterprise with new significance. Although numerous in number, such enterprises are still small in scale and are concentrated mostly in retail business and services trade.

(4) **Joint enterprises**—These enterprises are jointly financed and established by different domestic economic organizations and by domestic organizations linking with foreign companies. Since 1978, with the transfer of economic decision-making power downwards, some of the original economic

organizations have tried to shift from a vertical form of economic connection to horizontal integration. There has appeared an unprecedented number of trans-sectoral and trans-regional economic organizations and those which have multiple owners (e.g., group companies). In addition, the process of economic reform in China is also one of opening to the outside world. In the development of this process, various economic organizations emerged in joint investment and co-production with foreign companies. They are concentrated mainly in the coastal provinces. These joint enterprises carry with them scale efficiency and have become a force which exerts an increasingly greater influence on the economic decision-making process.

## 4. The Present Situation

Since 1978 economic power decentralization has been conducted along two tracks. On one track, power decentralization is conducted within the original social and economic organization from top to bottom. It started in the field of distribution and spread gradually to other fields. Local governments have enjoyed increasing power in output planning, planning of material distribution, investment approval, use of foreign exchange, balance of exports and imports, use of foreign investment and planning of credit. On the other track, the decentralization of power has produced new decision-making bodies both inside and outside of the original system. Economic decision-making power is gradually being transferred to the people who run the enterprises. This transfer is indicated by the increasing proportion of product sold by the enterprises, the increasing proportion of raw materials purchased by the enterprises, the extension of the power of enterprises to employ or dismiss people, the extension of their power to choose the kinds of products they produce and to choose investment direction. This means that enterprises are now becoming real enterprises. The planning system is turning towards a market system.

These developments are indications of the profound changes that have occurred in the planning and decision-making system. The appearance of new decision-making bodies shows that the principles, mechanism and means of economic decision-making are experiencing fundamental change, namely, the system of direct control of the economy is evolving into a system of indirect control. These changes are manifested in the following aspects.

(1) The division of responsibility between the party and government has become clear. In the traditional system, the party meant the government and the government meant the party. As economic reform deepens, government has become increasingly more important in decision-making and the party has become more of a political organization, no longer unduly intervening in the daily decision-making of enterprises.

(2) The economic decision-making of the government and the enterprises has been separated, with governments no longer intervening in the daily decision-making of the enterprises. In the traditional system, government not only managed the macro-economy, but, as owner of the enterprise's assets, made its economic decisions. As a result, in a micro sense, the responsibility of the government and the enterprises was not separate and the enterprises existed only as government factories and were not responsible for increasing the value of their assets. At the macro level, taxes and profits also lacked separation with government revenue coming mainly from profits. But it is difficult to examine the profits obtained from the assets and, as a result, the so-called "big pot" phenomenon appeared, reducing the efficiency of the enterprise. As economic reform deepens, enterprises have begun to have more independence and government is compelled to conduct economic decision-making as manager of the macro-economy.

(3) Decision-making on social and economic issues has also begun to separate. In the tradition system, they were integrated and an organization in charge of economic activities was also in charge of social activities. As economic reform deepens, there is a trend towards the separation of these two functions. This trend is manifested not only in developments relating to unemployment assistance, health benefits, pension funds, higher education expenses and commercialization of housing, but also in the development and changing functions of many social organizations representing different social strata and groups. Such organizations as the Association of Enterprise Managers and the Association of the Aged have appeared. Trade unions, the Youth

League and the Association of Women, which already existed, have come to act more in the nature of social organizations.

The above mentioned changes have created an existing economic decision-making system with the following characteristics.

(1) The government has begun to respect economic law when making economic decisions. It no longer employs administrative orders and political mobilization in the pursuit of economic construction. Instead, it has begun to use interest rates, tax rates, exchange rates and prices to manage the macro-economy. This indicates that the Ministry of Finance and the People's Bank of China can play a greater role and that the State Planning Commission must use new measures in decisions affecting the macro-economy. Meanwhile, maturing markets enable the government to focus on measures to make it more effective, including strengthening sectors in charge of macro-economic management such as statistics, taxation, national economic accounting and the management of state-owned assets, as well as reducing the role of the specialized industrial ministries. Nurturing markets, improving the function of government, and establishing a new system of macro-economy management must constitute the main work for the State Planning Commission.

(2) Decision-making power has decentralized as the power and scope of the central government has decreased. Before 1978, the State Planning Commission had 256 kinds of products under its control, while at present it has only 20.[5] At the same time there has been a corresponding increase in the economic decision-making power of individuals and junior levels of government.

(3) As power decentralization becomes more wide spread, more technical knowledge is required in decision-making and a large number of different kinds of advisory and research bodies have appeared. Before 1978, there were only a few research bodies, most of whom were engaged in theoretical study. Little research work was done on applied economic theory and matters related to decision-making. But as economic reform took off, a large number of research institutes and think-tanks have developed to serve economic organizations.

(4) The systematized means of managing a decentralized economy have begun to take shape with the strengthening of the power of the National People's Congress to issue economic legislation and the increased capability of judicial bodies to deal

with economic cases. By November 1988, the National People's Congress had published 39 pieces of economic legislation in a period of ten years, accounting for 54 percent of all the legislation published during that period. Economic legislation passed by the People's Congresses at provincial levels also grew in number. They totalled 485 by 1987, which amount to 1/2 of the total provincial legislation. Meanwhile, the courts in the country receive an increasing number of economic cases, from 44,000 in 1983 to 366,000 in 1987.

(5) Finally, as the reform goes on, different interest groups have appeared, which begin to exert influence on decision-making, as evident by the role played by China's mass media.

The above-mentioned characteristics show that since 1978, China's economic policy has experienced profound changes. The decentralization of decision-making power requires that decision-making be scientific and enjoy systematized guarantees. The advisory bodies for decision-making have increased. The legal system is amplified and public opinion plays a supervisory role. These changes have all improved transparency and, to a certain extent, the quality of decisions.

## 5. Problems and Trends

Reform and power decentralization in socialist countries can follow two models. One is administrative power decentralization and the other is economic power decentralization.

When analysing power decentralization in China during the period from 1978 to 1988, we find that it has both administrative and economic characteristics. The principle of the market and that of planning are practised simultaneously in what has been described by Chinese economists as a "double-track" economic system. This dual nature is clearly shown in the fact that every product (including labour and capital) carries two prices. One is the list price fixed by the government (the planned price). The other is the floating price determined by market forces or by both parties in the business (the market price). The so-called planned price means that the enterprises, especially state-owned enterprises, must fulfil the production target set by the plan and products are priced, allocated and distributed by state-controlled channels. New enterprises outside the original system and output in excess of the production quota of the state-owned enterprises are governed by the market price which means that enterprises are responsible for the pricing and marketing of their products and the purchasing of raw materials according to the principle of market.[6] The coexistence of the two tracks indicates that both the input and output of the enterprises face two prices at the same time. Generally speaking, market prices are much higher than planned prices. In the iron and steel industries for example, the market price is two or three times higher than the planned price.

The coexistence of the principle of market embodied in economic power decentralization and the principle of planning embodied in administrative power decentralization[7] has produced this double-track system which in turn has produced some special problems for economic decision-making.

1. The lack of a unified price has resulted in ambiguous, distorted production signals. Distorted signals result in improper allocation of resources which in turn further distort signals resulting in production fluctuations. Because of ambiguous signals, technical input into the decision-making process has increased drastically and enterprises have turned to think-tanks for help in signal-recognition so as to maximize profits.

2. During the ten years of economic power decentralization independent interests have emerged who wish to have the advantage of both tracks and to avoid each of their disadvantages. For instance, because of low prices all enterprises wish to have their input allocated according to the principle of planning. Meanwhile they want to be in charge of the marketing of their own products according to market principles because of high prices. They often vacillate between the two tracks, acting now on the principle of market and later on the principle of planning, and there is a constant state of negotiations with the government. Management and local governments in particular have responded by highlighting the principle of market one moment and planning the next. They even go so far as to stress both these principles simultaneously. As a result, various decision-making bodies bargain with each other without any regulation and there is a lack of decision-making standardization.

3. The coexistence of two tracks has produced ambiguous rules under which decision-making bodies gradually grow

indifferent to long-term economic considerations, focusing only on the short term. Enterprises have become greatly concerned with profit but not with responsibility that should accompany economic loss.

It is evident that this situation is not conducive to the development of China's economy. What's more, such unsystematic and unstandardized decision-making will bring about social and political problems. China's government has realized this point and proposed that a new order of socialist commodity economy be established.

## 6. Reform Trends

The question facing China's economy is not whether it needs decentralization but whether administrative decentralization should be brought into the orbit of economic decentralization or vice versa.

In many respects the recent reforms have strengthened economic decentralization. First, the position and the role of individuals in social and economic life have been realized and indeed is being confirmed. The general principle of China's reform is that individuals enjoy relevant economic rights and are accountable for the results of their economic activities. With regard to the income at their disposal, individuals have the right to buy consumer goods of their own choice and can use their income for direct or indirect investment. While making painstaking efforts to increase the production of consumer goods, China has also opened stock markets and bond markets for people to invest in. With respect to the choice of profession, labour and service markets have appeared. Individuals now have the right to choose their profession. Accordingly employment opportunities and wage levels are beginning to be governed by markets. In the field of social welfare, China has begun to reform some of the old practices in which everything was run by the central planning agencies. In the area of tax reform, individuals have begun to experience a rather comprehensive monetization—for example, the commercialization of housing and the establishment of a fee-for-service system of medical care.

Second, the status and role of enterprises as the main bodies of economic activity have also been realized and confirmed. The general principle of reform is to change all enterprises (state-owned, collective or private) into organizations which can manage their economic commodity activities independently and benefit from profits while being held accountable for losses. Enterprises should have ample freedom of action so long as they obey the nation's laws. China is already engaged in the work of introducing stock in enterprises and selling or renting small enterprises to individuals. At the same time, the bankruptcy system has been introduced to state-owned enterprises.

Third, it has been realized that the invisible hand should complement the visible hand. In this connection, the general principle of reform is that governments should only interfere in fields where the role of the invisible hand is inappropriate. In fact, the government is already reducing its interference and replacing the system of prices fixed by planning with the one of prices determined by markets. Labour, services and finance markets are beginning to take shape. The tendency is for government to be responsible only for investment in infrastructural installations and facilities or services beneficial to the public.

Guided by the above trends, power decentralization in China's economy is moving from the purely administrative decentralization to economic decentralization. With the development of markets, the central government has weakened and disbanded some specialized ministries. There has appeared a gradual transition from a tax sharing system to a system of tax categorization on which a standardized governmental management system capable of adapting to markets is established at different levels. Meanwhile, in the light of their respective economic situations, local governments at different levels have also experienced functional changes.

In short, through ten years of reform, people have come to realize the value of reform oriented markets. Meanwhile, different social strata and bodies have appeared. As the process of reform develops, decision-making in China's economy will be more democratic and scientific.

## Notes

1. Chinese economists hold different views in their understanding of the objective of China's reform. The core of their differences is whether it's planning based on the markets or the markets based on planning. But they have one thing in common. They all recognize that market mechanisms must be introduced into China's economy.

2. Table 1 is based on: Zhou Tai-he, *Contemporary Economic System In China*, (China's Social and Scientific Publishing House, 1984); Liu Ji-ruen, "On Administrative Power Decentralization" contained in *Comparison of Social Economic Systems*, 3 (1988); Huang Chu-hua, "The Role of China's Local Government in Planning and Markets", *Comparison of Social Economic Systems*, 1 (1987).

3. Research Institute of China's Economic Structural Reform, "An Investigation and Preliminary Analysis of The Qualities of The Cadres in the Enterprises in Contemporary China", contained in *Reform: The Challenge and Choice Before Us*, (China's Economic Publishing House, April 1986), p. 16, pp. 270-305.

4. Research Institute of China's Economic Structural Reform, "The Social-psychological Reflection of the Reform of Prices", *Reform: The Challenge and Choice Before Us*, (China's Economic Publishing House, April 1986), pp. 71-90.

5. The so-called products under planned control are those whose raw material supply, marketing and pricing are fixed by the government.

6. The output in excess of production target means that the country gives a proportion of self-marketing and self-purchasing to its enterprises. For example, in iron and steel industry, the proportion is 20 percent.

7. Administrative decentralization is only an improvement on a highly centralized economy. It should not be considered in excess of the principle of planned economy.

# References

The Bureau of Integrated Planning of the Committee of National Economic Structural Reform. *The General Train of Thought of China's Reform.* (Shen Yang Publishing House, 1988).

Gruchy, Allan G. *Comparative Economic Systems.* (Houghton Mifflin Company, 1980).

*The Economic Structural Reform in the USSR and the Socialist Countries in Eastern Europe.* (Chongqing Publishing House, 1986).

The Research Institute of China's Economic Structural Reform, *Reform: The Challenge and Choice Before Us.* (China's Economy Publishing House, 1986).

The Research Institute of Economy, *China's Economic Structural Reform.* (China's Economy Publishing House, 1987).

Wu Jing-lian and Zhou Xiao-chuan. *A Collection of the Planning of China's Economic Structural Reform: From 1979 to 1987.* (Prospects of China Publishing House, 1988).

Wu Jing-lian and Zhou Xiao-chuan. *The General Planning of China's Economic Structural Reform.* (Prospects of China Publishing House, 1988).

Zheng Tai-he. *The Economic Structural System in Contemporary China.* (China's Social and Scientific Publishing House, 1984).

# Hong Kong

## Economic Policy under Stress

*H.C. Kuan*

## Introduction

This paper aims to examine the circumstances under which economic policies in Hong Kong are formed. While basic, more stable factors are important and will be addressed, the emphasis will be laid on the changing conditions. The scheduled return of sovereignty over Hong Kong to China in 1997 has irrevocably changed the political landscape of this once tranquil British colony. The issue of governability in the 1990s represents a scissoring effect of increasing demands and decreasing supports. While previously the government of Hong Kong had to serve only one master, it must now satisfy both the British and the Chinese governments. In addition, the recent politicization of the society has awakened the Hong Kong Chinese from their political hibernation and they are becoming more demanding of their government. On the other hand, however, the government's authority is declining and it is increasingly looked upon as a lame duck administration, while the citizens' commitment to Hong Kong has become uncertain, as attested to by the tide of brain drain.

## The Ecology of Economic Policy

The most constant factor that underlies the policies of the government is the nature of the economy. It is small, open and externally-oriented. In the absence of natural resources, Hong Kong has to import virtually all of what it needs and concomitantly has to export to meet the costs of imported goods. The economy of Hong Kong has been market-regulated, with a free flow of goods and capital. Therefore, it is highly exposed to cyclical movements in the external markets.

Such a small, open and externally-oriented economy leaves little scope for the government to manipulate, much less to plan. Economic life in Hong Kong is largely determined by thousands of decisions made by individual and corporate players coordinated by the market. The most the government can do is to play a supportive role. The structural transformation of the Hong Kong economy over the past decades—the flourishing entrepot trade after the Second World War, its end and the growth of manufacturing after the Korean War, and more recently the expansion in financial services—has not come about as a result of government economic policy.

## Paradigm for Policy-making

The second important factor that governs economic decision-making is the traditional principle of minimum government intervention.

With quite a remarkable persistence, the government of Hong Kong has subscribed to the idea of limited governance with regard to economic life. The basic policy of non-intervention was eloquently defended by Mr. (later Sir) John Cowperthwaite, the financial secretary from 1961-1971 as follows:

> I suppose it is inevitable that Government should be regarded as wrong if it interferes when in public eyes all is apparently well... and equally wrong if it does not interfere and later things go bad. But, in any case, I largely agree with those that hold that Government should not in general interfere with the course of the economy merely on the strength of its own commercial judgement. If we cannot rely on the judgement of

individual businessmen, taking their own risks, we have no future anyway.... I still believe that, in the long run, the aggregate of the decisions of individual businessmen, exercising individual judgement in a free economy, even if often mistaken, is likely to do less harm than the centralized decisions of a government, and certainly the harm is likely to be counteracted faster.[1]

Under the policy paradigm of non-intervention, private efforts have been the driving force of the economy of Hong Kong, whereas the role of the government has been largely confined to providing the infrastructure, such as maintaining a free and open trade system, ensuring the adequate supply of energy, water and land for the industry, and strengthening the legal and institutional framework in which commercial and financial institutions operate.

The specific implications of the paradigm of minimum interference are manifold. For instance, we may argue that the Hong Kong government has no industrial policy, if by this it is meant the strategic industry approach adopted by Taiwan and South Korea. In Hong Kong, no particular industry has been subsidized. In the old days, even vocational training was not considered to be worthy of support by general revenue. In labour policy, the government has refused to legislate for minimum wage protection. No income redistribution policy has been pursued, although government actions in providing public housing for the poor did have an unintended redistribution effect. As a sacred principle, the income and corporate tax rate has been kept low to encourage entrepreneurs to invest and employees to work hard. It is believed that funds left in the hands of the public will better generate growth and funds for the exchequer to use in the future.

Equally sacrosanct has been the principle of "do not use a fiscal policy to stimulate the economy." It is believed, that given the nature of the Hong Kong economy, deficit financing cannot stimulate growth. Since Hong Kong can not produce more than a small fraction of what she consumes, increased consumption stimulated by government deficits would mean increased imports without matching exports, thereby resulting in balance-of-payment crisis. Thus, the government has always tried to ensure that the rate of growth of public sector expenditure should not exceed that of the gross domestic product, that there is a broad balance of revenue and expenditure, and that at least half of the capital expenditure can be financed from the operating surplus.

Of equally remarkable stability is the government's monetary policy. The role of government has been confined to setting monetary standards and regulating financial institutions: Hong Kong has no central bank; its dollar is fully convertible; and there is no foreign exchange control. Two commercial banks, the Hong Kong and Shanghai Banking Corporation, and the Chartered Bank, are authorized to issue banknotes against a backup system. The money supply is largely determined by balance-of-payments performances. The only means of monetary control is the Interest Rate Agreement which was established in 1964 after an interest-rate war among the licensed banks. Since then, maximum interest rates are set by the Hong Kong Association of Banks on deposits of a range of value and of original maturities up to a specified period. On the determination of these rates, the Association is statutorily obliged to consult the government so that the wider public interest, such as the stability of the exchange rate, will be taken care of.

The role of the government in economic development has been a constant subject of public debate. A scrutiny of the proceedings of the Legislative Council in the 1960s and 1970s reveals a regular demand on the part of the Unofficial Members for more government involvement in the economy.[2] They were joined by academics who argued that the nature of Hong Kong's economy did allow a more active role of the government.[3]

Financial secretaries after Mr. Cowperthwaite have adopted a more positive stand in economic involvement. In the words of former financial secretary, Philip Haddon-Cave, the government followed a policy of "positive non-intervention" which could be defined as the view that:

> ... in the great majority of circumstances it is futile and damaging to the growth rate of the economy for attempts to be made to plan the allocation of resources available to the private sector and to frustrate the operation of market forces which, in an open economy, are difficult enough to predict, let alone to control.[4]

Beginning with Haddon-Cave, the government has become more active. The establishment of the Advisory Committee on

Diversification in 1977, and its report in 1979, represents a landmark in the economic history of Hong Kong. Since then, the government has taken on increasingly more responsibilities such as assisting in trade promotion and ensuring access to overseas markets, subsidizing vocational training, and providing support services for improving product quality. It is interesting to note that notwithstanding these changes, the government still claims that it has been pursuing the policy of positive non-intervention.

## The Basic Structure of Decision-Making

Politics in Hong Kong has been conceptualized in various ways, ranging from "administrative absorption of politics"[5], through "administrative, no-party state"[6] to "bureaucratic polity"[7]. Central to all these conceptualizations is the dominance of the bureaucratic government in the decision-making process. In the absence of party and electoral politics, the bureaucracy is highly autonomous. The Legislative Council, as a weak "parliament" without a representational role, and public opinions, which are often divided, can offer no effective checks against the bureaucracy. Although there are powerful interest groups, with some of them even institutionalized into the consultative process, no group has yet succeeded to capture the bureaucracy to serve its interests. The secured tenure of the government and the autonomy of the bureaucracy has thus far enabled the government of Hong Kong to define its role in its own way, thereby upholding the myth of positive non-intervention.

Furthermore, the absence of intermediate institutions to aggregate the diverse interests of the society has left a vacuum to be filled by the government who can claim the ultimate right to judge on the general interest. A recent dialogue between a journalist and a high-ranking official, though referring to the information policy of the government only, is illustrative of the perception of the government.

    Journalist:  Don't you think that the public has right to know?
    Official:  I certainly believe in that, but the right to know is not without restriction.

| | |
|---|---|
| Journalist: | Who decides on how much restriction then? |
| Official: | The restriction must be in the public interest. |
| Journalist: | Who decides on what is the public interest? |
| Official: | The Government... |
| Journalist: | The Government, as it is well known, is not impartial. It may hide facts in order to protect itself. |
| Official: | The Government of Hong Kong differs from other governments; this government can not be brought down. It does not have to protect itself in this way.[8] |

However, while the bureaucracy is dominant in policy-making, it is sensitive to public opinion. It is a constant feature of bureaucratic rule in Hong Kong that no attempt is made to introduce or change a policy without consultation with interested parties. Subject to the rules of the game, as largely defined by the bureaucracy, political influence can be informal or institutionalized, effective or futile. Like many other countries, key actors in this process of influence are, as far as economic policies are concerned, leading business and industrial associations and labour groups.

Strategic elites and interest groups enjoy easy access to the power centre and may achieve what they want at private meetings or through formalized representation. Key players in this process are the Federation of Hong Kong Industries, the Hong Kong General Chamber of Commerce, The Chinese Manufacturers' Association of Hong Kong, The Hong Kong Tourist Association, and the Hong Kong Management Association. For instance, the quota system for the textile industry was initiated by the Federation of Hong Kong Industries and then accepted by the government in 1963. As the Federation is, on the whole, representative of big industries, it is no wonder that the system tends to protect their interests at the expense of their smaller counterparts.[9]

Influence by labour compares unfavourably with that of business, because the former are ineffectively organized. The prominence of small-scale production in Hong Kong, the horizontal mobility of workers, the political divisions among trade unions and the restrictive laws governing union amalgamation and sharing of leadership all contribute to a weak union movement in Hong Kong. As a result, labour interests tend to be better advocated by cause groups such as the Christian Industrial Committee, which have often resorted to public campaigns or unconventional pressure activities to move the government bureaucracy.

The unequal access to the power centre has been lamented in the literature from time to time.[10] Davies even posits a closely knit structure of ruling elites with similar ideologies, interests, and living style,[11] which smacks of a conspiracy of the rich against the poor. It is, however, difficult to document how the ruling elite model works in reality.

In contrast to the informal process of influence by elite networks, the advisory system is a visible and respected structure of influence. There are more than three hundred, statutory or non-statutory, permanent or ad hoc advisory committees and boards which cover a wide variety of policy areas. The relative importance of advisory committees depends on many factors. In general, one may judge a committee's influence by noting its chairperson and composition. For example, the Industry Development Board, chaired by the Financial Secretary, is a powerful one. Other prominent standing advisory committees for economic policies are: the Trade Advisory Board and the Textile Advisory Board, chaired by the Secretary for Trade and Industry; the Labour Advisory Board, with the Commissioner for Labour as the ex-officio chairman; the Land and Buildings Advisory Board, with a chairman from the private sector; and the Transport Advisory Committee, with a Legislative Councillor serving as the chairlady.

The advisory system is an important linkage between the government and society. It provides the bureaucracy with much needed information, expertise, legitimacy and support. It assures the society of an institutionalized channel of influence. However, there is as yet no systematic study on its workings.

## The Decision Process

The initiation of the decision process in Hong Kong can be characterized as "bubble-up". As the policy paradigm of positive non-intervention implies, the government is not an eager initiator. The policy of moving industry out of domestic-type buildings into factory premises in the late 1940s and the proposal to amalgamate the four stock exchanges in the late 1960s are notable exceptions. The government prefers to let business and industry fend for themselves. Increasing involvement of the government in economic regulation or promotion has been the result of crises such as the banking crisis in the mid-1980s, or of relentless demands from the society for assistance, such as the case of export risk insurance in 1966.

After initiation, a process of exploration is undertaken in which the decisional situation is defined, values and objectives studied or reviewed, and policy alternatives generated and evaluated. The realities are complex and vary from policy to policy. Under normal circumstances, a relevant department or policy branch of the Government Secretariat has to conduct a study and come up with a policy paper after having sought the views from a standing advisory committee and from the parties affected. Transportation and labour policies are often formulated in this way. For more technical issues such as the control of water pollution, a professional consultancy report may be engaged. On a major issue, initiative may be entrusted to an ad hoc advisory board specially set up and led by a policy secretary or an influential Legislative Councillor. A prominent example is the Special Committee on Land Production established in 1977. Where popular support to the proposed policy is regarded as desirable, an exercise in public opinion gathering may be mounted through the city district offices, or in the form of a Green Paper, although no economic policy has ever been explored in this way.

Policy proposals involving expenditure have to go through the scrutiny of the Finance Branch before they can be presented to the Executive Council for decision. This requirement confers upon the Financial Secretary a more powerful position vis-à-vis the policy secretaries. In this sense, policies are dictated by financial considerations. In 1968, the decision to build the Mass Transit Railway was effectively objected to by the financial secretary, Mr.

Cowperthwaite.[12] A recent proposal to stage the World Exposition in 1997, as a political confidence booster, was also rejected on financial grounds.

A policy decided upon by the Executive Council needs the support of the Legislative Council, especially when legislation or supplementary budget is required. In the old days, this support was rarely withheld.

## Challenges in an Era of Changes

In the 1980s, Hong Kong stands at a crossroads because there are critical challenges both to the economy and the political system.

The challenges to the economy of Hong Kong are well known. At the global level, international trade has been moving away from free trade, with a sustained threat of increasing protectionism in the United States and with the emergence of new trading blocs such as the European Economic Community in 1992. At the regional level, Hong Kong faces keen competition both from technology-intensive producers in Taiwan and Korea and low-cost industries in Thailand and Malaysia. To weather the storms, the economy has to be restructured.

The challenge to the domestic politics of Hong Kong have been succinctly analysed by Lau.[13] In a nutshell, it refers to the departing government's unenviable job of undertaking reforms that will serve as a foundation for a future government. In the process, these measures have contributed to a weakening, instead of a strengthening, of the authority of the incumbent government. With its authority diminishing year after year, the departing government can hardly have the will to carry out effective policies to assist in the restructuring of the economy.

The challenge to the governability of Hong Kong has an external dimension too. It is the "China factor" which may prove itself as the most important factor in the economic development of Hong Kong. The China factor is a complicated set of parameters related to the issue of 1997 and to the increasing interdependence between China and Hong Kong.

First, the "open policy" of China has contributed substantially to the changes in the economic environment of Hong Kong, thereby offering decision-makers various options for development.

Second, the planned transfer of sovereignty of Hong Kong to China in 1997 and the unsettled business of political transition have affected the temporal parameters, according to which the Hong Kong government or private entrepreneurs make up their minds on planning and investment. In the words of Howe, the danger is "that if the 1997 focus becomes excessive, local institutions and policies will freeze in the interim period."[14] Fortunately, the period of freeze has begun to give way to a thaw, although the 1997 focus still lingers on. Third, the requirements of Sino-British cooperation in making arrangements in the transitional period for the smooth transfer of government in 1997, as well as the unfolding of economic influences from China, have made the Chinese government a significant factor which has to be taken into account in any major economic decisions in Hong Kong.

The opening up of China was initially conceived by many in Hong Kong as a grave challenge of yet another low-cost competitor in the world market. As it has turned out, however, investment opportunities created in China have allowed more and more Hong Kong factories to move their operations to China to take advantage of low labour and land costs. It is estimated that over 70 percent of foreign investments in China have been made by Hong Kong firms and more than four million workers in China are employed by Hong Kong manufacturers. At home, industrial employment in 1987 was 875,250. Thus, the farming out of production to China has postponed the pressures for the Hong Kong industry to move up-market based on high technology.

The flourishing trade with China is also changing the vista of the Hong Kong economy. The trend of economic development in Hong Kong has been toward greater dependence on trade. The total value of visible imports and exports in Hong Kong rose to about twice the size of the gross domestic product, from around 1.5 times in the early 1960s. As trading opportunities in industrialized countries deteriorate as a result of recession and protectionism, China suddenly came to the rescue. Since 1985, China has been Hong Kong's largest trading partner[15]. In 1987, China is the number one supplier of Hong Kong's imports, providing 31 percent of the total, and the number two receiver of Hong Kong's exports, claiming 14 percent of the whole. Changes in the pattern of re-exports are even more significant. Hong Kong's re-exports in the said year accounted for 48 percent of the combined total of domestic exports and re-exports, with China being the largest country of origin and destination. In a word, Hong Kong has resumed its traditional role of entrepôt trade for China.

Thus, the economy of Hong Kong has been relieved of much of its stress under fierce world competition due to the restructuring of manufacturing industry through relocation of production and to the prosperity in China-related re-exports. Can Hong Kong continue to survive on this basis? Some businessmen are optimistic. But for some entrepreneurs, the present opportunities are only buying time for the manufacturing industries. As manufacturing is still the backbone of the Hong Kong economy, Hong Kong must, in the long run, move toward more design intensive and knowledge based manufacturing. Any up-market movement of the economy is inconceivable without strategic planning and long-term investment, which in turn can not be contemplated without taking the political future of Hong Kong into account. The issue of 1997 has therefore been quite disturbing, to say the least. In general, the 1984 Agreement between Britain and China on the return of sovereignty of Hong Kong to China in 1997 represents an important step which has somewhat cleared the way for the further development of the economy. Compared to the period of 1982 through 1984, the climate for investment has been improved. Still, there are sufficient uncertainties to prevent long-term investment from being ventured. For many players, the economy of Hong Kong is nourished by short-term, renewable confidence.

In another sense, the Sino-British agreement has frozen the economic system, which may further constrain the Hong Kong government. To alleviate fear of Chinese interference, Article 3(5) of the Agreement declares that the current economic system in Hong Kong will remain unchanged; Annex I(VI) provides that the future Special Administrative Region shall maintain the capitalist economic and trade systems "previously" practised in Hong Kong, and there are other annexes pertaining to the maintenance of "previous" systems or practices in the monetary and financial systems, shipping management and regulation, and civil aviation management.

Whether this guarantee of preservation of the status quo implies restriction of innovation in economic policies depends on the meaning of the word "previous" (whether it refers to the date of the signing of the Agreement or the date of transfer of sovereignty) or on how the Chinese government would extemporaneously react from case to case. Either way, the Chinese government has become a significant player in the economic system of Hong Kong.

The role of the Chinese government is absolutely certain in the area of external economic relations. To preserve its right to enter into commercial agreements and to negotiate and renew them with foreign states and international organizations, Hong Kong has no choice but to depend on the participation and support of China.

Turning to the domestic scene, the Sino-British Agreement has formally institutionalized China's participation in policy-making on matters of land, which is scarce and significant for economic development. Annex III of the Agreement specifically prescribes that as from the entry into force of the Agreement, the granting of new land shall be limited to 50 hectares a year, that the income obtained by the Hong Kong government from land transactions shall be equally shared between the former and the future Special Administrative Region government and that the share of the latter government shall be deposited in banks and shall not be drawn on except for projects approved by a Land Commission composed of an equal number of officials designated by the British and the Chinese government.

The Sino-British Joint Liaison Group set up under the Sino-British Agreement provides yet another channel for Chinese participation in economic decision-making in Hong Kong. Although the Group was established with the primary aim of facilitating the transfer of government in 1997, it can, however, "exchange information and conduct consultations on such subjects as may be agreed by the two sides."

The general point permeating all these arrangements is that the consent of the Chinese government is required whenever a policy measure is expected to bear consequences beyond 1997. This spirit can be extended to areas not specified in the Agreement, for practical reasons, if not for moral obligation.

The China factor does not only affect the "constitutional" framework for economic decision-making, but also manifests itself as an emerging economic force. China-owned economic units established in Hong Kong are fast expanding and capturing influence in the market.

Before 1979, China-owned business organizations wielded little power in Hong Kong's economy, preferring a rather dormant role. The open policy of China has encouraged an aggressive expansion of these Chinese establishments. The general pattern is towards corporate amalgamation, managerial reforms, and assertive business practices. Suffice here to use the Bank of China as an example. Since the 1979 conference on overseas branches of the Bank of China, 13 China-owned banks have been integrated under the Bank of China and have become the Bank of China Group. In 1983, the Group was divorced from the People's Bank of China and placed directly under the State Council of the Chinese government. In 1985, the Group commanded a total asset of HK$90 billion, making up 7.5 percent of the total banking assets in Hong Kong. With a total of HK$60 billion in deposits, the Group claimed, in the same year, 15 percent of the total bank deposits in Hong Kong.

In addition to the old China-owned business organizations, many new ones have been established since 1979. Some of them came with powerful backing by national and provincial authorities in China. A prominent example is the Everbright Corporation. The burgeoning of Chinese economic activity has significant implications for the economy of Hong Kong. First, it will affect the distribution of power in the market of Hong Kong. In this sense, 1986 was a watershed, when the Bank of China participated in the rescue of the troubled Ka Wah Bank. Second, since all but a few (the few are "private enterprises with government capital") of Chinese business groups are government agents, there is a natural tendency for the Chinese government to get more involved in order to put the operations of these units under control. As closer political and economic ties grow, there may be the danger for China-owned enterprises to collude instead of playing by the rules of the game.[16]

## Governability, Choice and Development

The discussions so far pertain to the suggestion that as the very foundations of Hong Kong's economy have been shifting, the base of authority of the Hong Kong government has also been eroding. Building on the growing interdependence with China, there are several possible directions of economic development for Hong Kong: high-technology manufacturing, entrêpot trade, and as an international financial centre. How a direction, or combination of directions, should be pursued involves fundamental choices with long-term implications. The level of vision and the magnitude of the task require that private efforts be supplemented by more government involvement than what the policy of positive non-intervention suggests. However, this government is now constrained by its counted lifespan, its public image as a lame duck, the requirement to consult an incoming authority and the upsurge of pressure politics. There are also signs that bureaucrats are demoralized and that departments are choosing to refrain from suggesting policy directions. The danger that looms on the horizon is that the government may lose the will or the ability to govern.

In the area of economic policies, the following pattern can be discerned. The government can still be resolute in tightening up regulations, especially when hard pressed by crises; it is, however, extremely cautious and hesitant when it comes to long-term commitments.

As an example of the former style, we may cite the tinkering with the capital market. The 1980s has been a decade of active financial regulations. Legislation was enacted on disclosure of shareholdings, on insurance, on the unification of the four stock exchanges, and on the three-tier banking system. Government's firmness in regulating the stock market was unprecedented. The four stock exchanges were against unification, but were forced into it by the 1980 legislation. The government's firmness was exhibited again, eight years later, when the unified Stock Exchange was "advised" to establish a management committee to replace its popularly elected general committee and some members of this committee were subjected to investigation by the Commission Against Corruption and were "requested" to relinquish their powers to manage the business of the Stock Exchange. The government threatened emergent legislation should the Stock Exchange refuse to cooperate.

As to the latter style, it is understandable that a government which has always subscribed to a limited role and now has counted years to rule can hardly commit itself to any long-term expensive projects. As a matter of fact, the dominant trend of policies is to hive off as many services as possible. Prime examples are housing and medicine, where rising costs for more and better services are expected. New projects such as the replacement airport are even more complex and demanding. The result is a protracted process of indeterminacy.

## The Case of the Replacement Airport

It goes without saying that a modern international airport is crucial to an externally-oriented economy. Hong Kong's international airport, Kai Tak, was built in 1928 on reclaimed land in an urban area, with only one runway. Since then, it has been considerably expanded. As early as 1974, when a taxiway was added to accommodate wide-bodied aircraft, the Director of Civil Aviation warned that the failure to put up a new airport could impose serious costs on the Hong Kong economy.[17] In that year, the Kai Tak airport had a record total of 8.6 million passengers and 100,721 metric tonnes of cargo freight. Subsequently, a feasibility study was commissioned to a consultancy firm which reported in 1976 that a replacement airport could be built at Cheklapkok in north Lantau at a cost of HK$370 billion. The idea was dismissed by the government on grounds of financial considerations.

In 1979, the issue of a replacement airport was taken up again. It was realized that the further expansion of Kai Tak airport beyond 1981 was seriously restricted by operational constraints imposed by the nature of the surrounding terrain, the close proximity of the urban areas and the lack of land. On the other hand, the powerful Advisory Committee on Diversification also recommended in 1979 that in order to benefit from the role of entrepôt trade with China, Hong Kong's ability to handle an ever-increasing volume of commercial cargo must be improved.[18]

From 1979 through 1982, a series of studies on the need for a new airport and its possible site were commissioned. The consensus was that there was a need, and that Cheklapkok was the most favoured site. A phased approach to the construction was recommended, with an initial outlay of roughly HK$37 billion expected. In early 1983, the government decided to postpone the decision on the replacement airport, on the following grounds:

1. the slowdown in the growth rate of the Hong Kong economy, especially for the manufacturing industry;
2. the rise in the interest rate;
3. the deficit of the government, being the first since 1974; and
4. the more urgent task of negotiating the multi-fibre agreement.[19]

The above reasons given by the then financial secretary were regarded by many as unconvincing. A more obvious cause was political, as the Sino-British negotiation over the future of Hong Kong had just started in a rather agitated atmosphere. The timing was conspicuously inappropriate for a major decision with an effect that would extend beyond 1997.

Interest on the issue of a replacement airport was then revived after the conclusion of the negotiation. In 1984, a total of 9.5 million passengers and 420,000 metric tonnes of freight passed through the Kai Tak Airport, representing almost 2.5 and 4 times the figures respectively in 1974. The need for a new airport was even more apparent. In 1985, China announced her intention to build an international airport in Shenzhen, the first Special Economic Zone of China, which was located just across the border from Hong Kong. On 22 January 1986, the Financial Secretary confirmed that the government was not actively planning for the construction of a replacement airport, but preferred to improve and extend the facilities at the Kai Tak Airport.[20] In mid-1986, a consortium composed of five major corporations in Hong Kong and led by Mr. Gordon Wu's Hopewell Holdings Ltd. proposed to build a new airport, together with port and road development, with the entire scheme to be financed by private interests (later known as the Hopewell Plan).[21]

It is interesting to note that Mr. Xu Jia-tun, Director of the New China News Agency, was reported to have told the consortium led by Hopewell that "China is solidly behind this (the Hopewell plan) project."[22] As reported in the same source, many mainland Chinese corporations also expressed interest in the Hopewell Plan.[23]

In response to the initiative of this consortium, a Steering Committee was set up under the Financial Secretary to study the Plan. The Chief Secretary met Mr. Wu in April 1987. Later, a Working Group was set up under the Strategic Planning Unit of the Lands and Works Branch of the Government Secretariat to further study the Hopewell Plan. From 1983 through the summer of 1987, the government's position on the idea of a replacement airport can be summarized, at best, as lukewarm. While the long-term need for a new airport was readily recognized, timing was regarded as crucial. In the meantime, it was decided to first extend the maximum useful life of the existing facilities which are judged to come under strain around 1995 and to reach saturation point around 2005. In the words of John Zaxley, then Acting Financial Secretary, "it is unlikely that a decision on a replacement airport needs to be taken before the fiscal year 1990-1991."[24]

Toward the end of 1987, the tone of the government seemed to have changed. In his first annual address to the Legislative Council on 7 October 1987, the new Governor, Sir David Wilson admitted that the Kai Tak airport "has limited potential for expansion and must eventually reach its ultimate capacity." Very interestingly, he remarked that "(e)ven before that point is reached we must take account of the environmental impact on Kowloon of an airport at Kai Tak developed to full capacity." He announced that the government has commissioned a joint Port and Airport Development Study and is expected to take some fundamental decisions in late summer of 1989 when the study will be completed.[25]

Within one month, the timetable was accelerated when the Financial Secretary informed the Legislative Council that "initial decision on the construction of a replacement airport could be taken towards the end of 1988."[26]

Thus, the evolution of the policy on the new airport is a protracted and indeterminate process. In the meantime, the air traffic has grown dramatically. In 1987, passenger traffic, at 12.7 million, was four times the 1974 level, and cargo carriage, at 611,000 tonnes, was six times the amount of that same year. Aircraft movements have increased 31 percent from 55,928 in 1979 to 73,370 in 1987. (There was no statistics on aircraft movements before 1979.) The strained capacity of the existing airport has been significantly alleviated by the introduction of wide-bodied airplanes. This source of relief is, however, drawing to exhaustion, as 80 percent of aircrafts visiting the airport in 1987 are already wide-bodied.

Domestically, there seem to be sufficient interest for a replacement airport. Within the government, The Civil Aviation Department has been, from the very beginning, a consistent supporter for a replacement airport. It has, however, failed to secure the support of the financial secretary so that the idea of a replacement airport has been guillotined three times on financial grounds. The financial considerations gradually gave away once cost-cutting construction technologies became available and the private sector vowed to take over the risks. The key factor shifted to the need to fit the airport development into an overall territorial development plan which could not proceed under uncertain political circumstances.

Outside the government, the opponents of a replacement airport are benefactors of the existing Kai Tak airport. Both Cathay Pacific and The Hong Kong Aircraft Engineering Company objected to the idea of a new airport due to the tremendous costs of relocating their service equipment. Though in the minority as far as social interests are concerned, Cathay Pacific has been an influential player in the policy of civil aviation and its views have to be seriously considered by the government.

Proponents in the debate are numerous. A new airport is clearly in the interest of the tourist industry. Both the Tourist Association and the Federation of Hong Kong Hotel Owners have supported the construction of a replacement airport. The issue has not been widely debated in the Legislative Council. Those who have spoken out are all in favour and some even complained of the sluggishness of the department in this regard.[27] The views of the environmentalists are ambivalent. In general, the location of an international airport in a crowded city like Hong Kong is always objectionable. If it is inevitable, it is then a matter of opting for the lesser evil. As an anomic group, environmentalists have long complained about the hazard of accidents and noise pollution inflicted on the residents of Kowloon City where the existing airport is located. The World Wildlife Fund, however, has objected to the construction of an airport near Deep Bay as proposed in the early 1970s, because it threatens to destroy the ecology of the Maipo Marshes, a natural reserve for transient birds, and the

Conservancy Association has opposed the construction of an airport on the Lantau Island, also in the interest of wildlife. The artificial island of the Hopewell Plan must have appeared to be the lesser evil. Their reaction to the newly proposed site of Lamma Island and Cheung Chau is not yet known.

Ultimately, it is the China factor that is of decisive importance. The indeterminacy of government with respect to the need of a replacement airport can largely be explained by the uncertainties surrounding the Sino-British negotiations, the pending clarification of issues related to Hong Kong's ability to enter into external commercial relations, especially in matters of civil aviation, and finally the general atmosphere in the relations between the Chinese and the British government. Since the signing of the Sino-British Agreement in 1984, clouds have begun to dissipate. There is hope again that the sky in Hong Kong can be reached via a new airport.

## Notes

1. *Hong Kong Hansard,* 1966, pp. 215-216.

2. *Hong Kong Hansard,* 1962, p. 72; 1963, p. 75; 1965, p. 162; 1967, p. 147; 1968, pp. 87, 125, 139-141; 1969, pp. 231-239 and 284; 1970-71, p. 43; 1972, p. 121; 1973, pp. 116, 593; 1974, pp. 58, 116.

3. See for instance, Nicholas C. Owen, "Economic Policy," in Keith Hopkins ed., *Hong Kong: The Industrial Colony,* (Hong Kong: Oxford University Press, 1971) pp. 141-206.

4. Sir Philip Haddon-Cave, "The Making of Some Aspects of Public Policy in Hong Kong," in David Lethbridge, *The Business Environment in Hong Kong,* (Hong Kong: Oxford University Press, 1980) p. xii.

5. Ambrose Y.C. King, "Administrative Absorption of Politics in Hong Kong: Emphasis on the Grass Roots Level," *Asian Survey,* No.5 (May 1975), pp. 422-39.

6. Peter Harris, *Hong Kong, A Study in Bureaucratic Politics,* (Hong Kong: Heinemann Asia, 1978) p.91.

7. Siu-kai Lau, *Society and Politics in Hong Kong,* (Hong Kong: The Chinese University Press, 1982) Ch.2.

8. Interview of John Chen, the Deputy Chief Secretary of the Hong Kong Government, by Margaret Ng, *Ming Pao*, 1 November 1988, p.7.

9. For an excellent analysis of the influence of the Federation of Hong Kong Industries, see Jörg Baumann, *Determinanten der Industriellen Entwicklung Hongkongs 1945-1979*, (Hamburg: Institut für Asiankunde, 1983) pp. 153-173.

10. Joe England, "Industrial Relations in Hong Kong," in Keith Hopkins ed., *Hong Kong, the Industrial Colony*, (Hong Kong: Oxford University Press, 1971) pp. 207-248; and N.J. Miners, *The Government and Politics of Hong Kong*, (Hong Kong: Oxford University Press, 1975) pp. 191-193.

11. S.N.G. Davies, "One Brand of Politics Rekindled," *Hong Kong Law Journal*, Vol. 7, No.1, pp. 44-84.

12. *Hong Kong Hansard 1968*, pp. 208-211. Finally, the government decided to go ahead with the Mass Transit Railway in 1973.

13. Lau Siu-kai, "The Unfinished Political Reforms of the Hong Kong Government," in John W. Langford & K.Lorne Brownsey, eds., *The Changing Shape of Government in the Asia-Pacific Region*, (Halifax, Nova Scotia: Institute for Research on Public Policy, 1988) pp. 43-82.

14. Christopher Howe, "Growth, Public Policy and Hong Kong's Economic Relationship with China," *China Quarterly*, No.95 (September 1983) pp. 532-533.

15. In 1978, China's share of Hong Kong's total exports was zero percent and that of total imports amounted to 16.8 percent only.

16. Sung Yun-wing, "The Role of the Government in the Future Industrial Development of Hong Kong," in Y.C.Yao et al., eds. *Strategies for the Future*, (Hong Kong: Centre of Asian Studies, Hong Kong University, 1985) p. 433.

17. "New $5,000M Airport Plan," *Hong Kong Standard*, 25 October 1974.

18. *Report of the Advisory Committee on Diversification*, (Hong Kong: Government Printer, 1979) p. 164.

19. "Cheklapkok Plan is Shelved," *South China Morning Post*, 26 February 1983, p.1.

20. Reply by the Financial Secretary in Legislative Council to the Question raised by Mr. S.L. Chen, *Hong Kong Hansard* 1985-1986, p. 506.

21. The five corporations are, besides Hopewell, Cheung Kong Holdings Ltd., Hutchison Wampoa, The Swire Group, and Jardine Matheson. The costs of the scheme were estimated to be around HK$25 billion, with one-fifth of it shouldered by the consortium and the rest financed by the Hong Kong and Shanghai Bank and the Citibank. Target date for completion of construction was scheduled for 1992 if government gave the go-ahead green light at the end of 1987. The consortium would charge toll fees for the use of the road network for thirty years, sell the marine berths and the new airport to the government either at cost or for a nominal sum in return for land development rights.

22. "Reach for the Skies," *South China Morning Post*, 14 June 1987, Sunday Spectrum, p. 1.

23. Ibid. The corporations involved are China International Trade and Industries Corporation, China Resources Ltd., The Guangdong Provincial Government, and the Bank of China.

24. Government Information Service News Release, "No Urgency in Deciding on Replacement Airport," 6 May 1987.

25. *Hong Kong Hansard*, 1987-88, p. 24. An Airport Development Studies Division was established in the Civil Aviation Department in October 1987.

26. *Hong Kong Hansard*, 1987-88, p. 376.

27. Question by Mr. S.L. Chen and Mr. Lee Yu-tai on 22 January 1986, *Hong Kong Hansard* 1985-86, pp. 505-506; speech by Mr. Lau Wong-fat on 19 March 1987 and on 5 November 1987, *Hong Kong Hansard*, 1986-1987, p. 1245 and *Hong Kong Hansard* 1987-88, p. 291; by Dr. Konrad Lam on 6 May 1987, *Hong Kong Hansard* 1986-1987, p. 1460; by Mr. Hilton Cheong-lean and Mr. H. Sohmen on 4 November 1987, *Hong Kong Hansard* 1987-1988, pp. 213 & 261 respectively; Mr. Edward Ho on 5 November 1987, *Hong Kong Hansard* 1987-1988, p. 297. Interests expressed in the Legislative Council are diverse. The majority spoke of the interests of industry and trade generally; some may have the interests of the New Territories in mind, where the new airport is likely to be located; still others would like to see the replacement airport as a booster for political confidence.

# Malaysia

## The Shaping of Economic Policy in a Multi-Ethnic Environment: The Malaysian Experience

*Mavis Puthucheary*

In multi-ethnic societies such as Malaysia, where competing interests are ethnically based, consensus is obtained through the formulation of "comprehensive" policies. In this way, the "open clash of organized interests", which is the basis for the formulation of policies in Western countries, is "conspicuously absent".[1] Public discussion on issues that are regarded as communally sensitive are seen as threatening to the stability of the system and therefore to be avoided. Instead, the key players in the decision-making process decide which set of policies in the comprehensive policy framework is to be given priority. It is therefore only at the implementation stage that one is able to determine the actual policies of the government.

Another way of discerning the central policies in the comprehensive policy framework, is by identifying the key players involved in the process. By identifying those involved in initiating the ideas that resulted in a change of policy and analysing the institutions and process through which the policy was formulated, one can understand why some policies are implemented with much greater sense of purpose and concentration of resources than others.

The purpose of this paper is to identify the key players and analyse the processes that resulted in the formulation of the New Economic Policy (NEP). The focus will be on the processes that resulted in the shaping of the central or "core" policies. That is, the policies that constituted the thrust of government's efforts, rather than the policies that were included at a later stage in order to get the desired political consensus. In the last part of the paper it is suggested that there has been changes in the emphasis given to certain policies, mainly resulting from changes in the values and orientations of the new set of key players in the policy-making process.

## Background to the NEP

The role of the government bureaucracy in policy formulation in Malaysia has been well documented. The special role of the administrative branch of the bureaucracy during the colonial period was described by Tilman in the following manner:

> During the colonial period in Malaya the administrative services—predominantly the Malayan Civil Service—constituted the predominant voice in the day-to-day control of affairs of government. The decisions of MCS officers, particularly those in the field, were likely to be highly discretionary; and MCS personnel had a hand in policy formulation on the federal level. Thus, the senior administrative service was often involved in decision-making that was sometimes as political as it was administrative, and this dominant position of the MCS in the political process elevated the Service and its members to a unique status comparable to that of the ICS in India.[2]

It was the colonial bureaucracy that decided to give preference to Malays in public employment and especially to open the elite MCS to Malays. A special school was established to educate the sons of the Malay aristocracy and a Malay Administration Service was set up to recruit and train Malays for eventual promotion to the MCS.

The decision to open the Malayan Civil Service to the sons of the Malay aristocracy was based on the belief that the British had a special responsibility to the Malays as the indigenous community whose Sultans had agreed to British intervention. Thus although the main decision-making power was clearly in British hands, there was an attempt to give at least a semblance of power to the Sultans by maintaining the structures of Malay traditional rule.

The opening of the MCS to Malays was also seen as another way to maintain the myth that power was shared with the traditional Malay elite. This was reflected in the attitude of the British officers to their Malay counterparts. There was little attempt to inculcate western values and norms of public interest and official conduct into the Malay civil servants. Instead, their traditional values of loyalty to the Sultans and to their group were strengthened by the kind of duties they were required to perform as civil servants. Their primary role was to work within their own community. Some of them even represented the Sultans and the Malay community in the appointed Federal Legislative Council.[3]

As the Malay middle class was concentrated in the civil service it was not surprising that the new Malay political elite that emerged after the Second World War was drawn primarily from the bureaucracy. A close relationship developed between the Malay bureaucracy and the United Malays National Organisation (UMNO), the party that claimed to represent Malay interests. UMNO's success in establishing a multi-racial government with itself as the senior partner, meant that the privileges given to Malays in public employment were preserved and strengthened in the post-independence period. In fact the "special rights" of the Malays became a fundamental part of the Constitution which gave the country its political independence. In particular, Malay officers in junior administrative positions took over the Malayan Civil Service, the key decision-making branch of the civil service. Furthermore, an ethnic quota for recruitment into this service guaranteed future Malay dominance in administrative decision-making. Thus, a Malay dominated administration "matched" a Malay dominated political structure.

At the time of independence the ownership of the economy was primarily in non-Malay hands. The foreign private sector owned about 60 percent of the economy, while the trade and service sector was in the hands of the Chinese, who owned small and medium-size businesses. There was an unwritten understanding that the government would not nationalize private industries.

Thus, during the first ten years of independence, economic policies tended to follow the pattern of the colonial period. Private investment in the economy was encouraged through various support schemes and incentives offered by the government. As in the past, these policies tended to benefit the urban, non-Malay sectors of the economy. In order to maintain and extend its support within the Malay community, the UMNO leadership focused its attention on the rural areas where the majority of Malays lived. Even before the country had gained its independence, UMNO leaders had pressured the colonial government to do more to improve the economic position of the Malays. This had led to the creation of special agencies, staffed mainly by Malay officers, to assist the Malay community. In 1952 the Rural and Industrial Development Authority (RIDA) was set up to provide credit and technical assistance to small-scale industries in the rural areas. Another agency to assist Malays was the Federal Land Development Authority (FELDA), a land development scheme to settle farmers in estate-managed land holdings.

Economic development in the first ten years of independence tended to follow a two-pronged approach. On the one hand, priority seemed to be on economic growth and especially on industrial development as a means to increase employment opportunities and reduce the country's dependence on the export of raw materials, particularly rubber and tin. On the other hand, the UMNO leadership emphasized the need to correct the economic imbalance between the Malay and non-Malay communities. Although government action to correct this imbalance necessarily meant the development of the subsistence agricultural sector through rural development programs, it was assumed that government action should eventually extend to cover a much wider scope including that of increasing Malay participation in the modern, urban sectors of the economy.

These two approaches to development constituted the economic policy of the government in the first ten years of independence. They represented the economic policy needs of the two major ethnic groups. Economic policy to promote economic growth was seen as benefitting the Chinese community, while rural development programs were seen as benefitting the Malay community. The importance of these two policies on the different races was reflected in the racial background of the political and administrative leadership. The two ministries that had the major responsibility for implementing the policies were the Ministry of Commerce and Industry and the Ministry of Rural Development.

Headed by a Chinese minister, the Ministry of Commerce and Industry was responsible for the implementation of policies designed to encourage private investment, particularly in certain industries. Tax incentives were given to industries which obtained pioneer status. The setting up of the Malayan Industrial Development Finance Limited (MIDFL) under the overall responsibility of the Ministry of Commerce and Industry, led to the establishment of industrial estates and the provision of credit and training for firms to expand their businesses. Later, a Federal Industrial Development Authority was set up to coordinate the activities of all organizations concerned with industrial promotion.

The Ministry of Rural Development, on the other hand, was established to achieve the politically desirable objective of improving the economic position of the Malay community. The two most important agencies responsible for providing economic assistance to Malays, FELDA and RIDA, came within the control of this Ministry. Its main function was to provide the machinery for the promotion of policies for assisting Malays, particularly those in the rural areas. The creation of this ministry indicated that the government's efforts were directed not only toward improving agricultural productivity—a function already being carried out by the Ministry of Agriculture—and represented its government's commitment to providing a wide range of assistance to the Malay community as part of the continuation of the special rights policy. In his study of rural development, Gayl Ness emphasizes this point when he states:

> The creation of the Ministry of Rural Development marked the major effort of the new indigenous leaders to take control of the administration they had inherited with independence and to use it to do the work of development as they saw it.[4]

The Ministry of Agriculture and the Ministry of Rural Development competed with each other for funds. At issue was not only the relative importance of the two ministries in improving the economic position of the Malays but also a personal struggle between the two ministers over who was to be the next prime minister.[5] By the early 1960s Tun Razak emerged the winner, and was soon able to establish an administrative network with which to expand his political base. Through the minor works projects under his ministry, Tun Razak was able to establish a patron-client relationship based on the distribution of rewards and benefits to those who were loyal to him, thus maintaining and even extending his popularity in the party. The rural development program, by its very nature, was much more broad-based than the other functionally-specific ministries. Rural development involved the provision of schools, electricity, roads, community centres, proper drainage facilities and mosques. The single most important ministry involved in rural development was the Ministry of Works. But according to Ness, the appointment of Tun Sambanthan as the Works Minister was an advantage to Tun Razak. As the head of the Malayan Indian Congress, the weakest party in the coalition government, Tun Sambanthan was not in any position to oppose the wishes of the future Prime Minister. Thus Tun Razak had "extensive control over the one ministry most crucial to the success of his program".[6]

By the time the Second Five-year Plan, 1961-1965, was drawn up, the budgetary allocation to the Ministry of Rural Development, including the allocations for schools in rural areas and other "Red Book" proposals incorporated in other ministries but under the control of the Ministry of Rural Development, was about equal to the allocation for the Ministry of Commerce and Industry.[7]

## The NEP Initiators

Although RIDA had been created as far back as 1952 to encourage the development of small-scale industries in the rural areas, there had been very little improvement in Malay participation in the commercial and industrial sectors. The main channel for upward mobility for the small but increasing Malay middle class was the government bureaucracy. But although there was an expansion of the government bureaucracy, particularly the MCS in the first ten years of independence, it was limited. Besides, it was becoming clear that salaries in the public sector could not keep pace with those in the buoyant private sector.

By the middle of the 1960s efforts were made to reactivate RIDA and particularly to extend the scope of its functions. A paper on its reorganization proposed that it should become directly involved in commercial and industrial activities instead of being content with providing small loans to small-scale entrepreneurs. It also proposed that the industries established under RIDA would be set up initially with the assistance of foreign experts and would employ and train Malays for management positions. In this way Malays would be able to participate not only in small-scale industry but also in large-scale business operations. It was assumed that these industries would be transferred to individual Malay entrepreneurs at a later date, but how this was to be done was not clearly laid down. The involvement of RIDA in industrial undertakings inevitably placed Tun Razak in potential conflict with the Ministry of Finance and the Ministry of Trade and Industry, both ministries headed by Chinese ministers. Tun Razak began to be convinced that efforts to improve the economic position of the Malays through increasing their opportunities to participate in commerce and industry, must be integrated into the whole development effort and not be regarded as a special program outside the main policy framework. In other words, efforts to involve Malays in commerce and industry required more than the setting up of the Ministry of Rural Development. It required a total commitment on the part of government. In the past, it had been found that policies to increase Malay participation in commercial and industrial activities had been opposed by Chinese ministers in the Cabinet who tended to construe these policies in terms of their effect on the Chinese community. This was seen in the early 1960s when the Minister of Agriculture tried to transform the private rice mills, largely owned by Chinese, into cooperatives. Pressure from the Chinese cabinet ministers resulted in a withdrawal of the proposal.[8] It was this attitude that forced RIDA to confine its activities to supporting Malay participation in industries which were not controlled by the Chinese. In a paper on the reorganization of RIDA, an officer criticized the former policy of his organization in the following way:

RIDA's activities were also influenced and circumscribed by the philosophy that we should not break other people's rice bowls. In other words we should not participate in economic activities in which non-Malays were already engaged. We should concentrate only on traditional crafts. This philosophy reduced RIDA activities to small, and in my opinion, unimportant sectors of the economy.[9]

The small group of government bureaucrats clustered around Tun Razak were convinced that there should be a complete reorientation of government policy so that the ministry in charge of promoting the economic development of the Malay community, which is how the they saw their role, should be in a position to tap the resources of the other ministries. In an effort to indicate this new orientation the name of the ministry was changed to the Ministry of National and Rural Development. But the change of name did not improve the status of the ministry in relation to others. The policy of the government continued to emphasize the dual set-up, with the Malay preference policies being largely outside the main policy framework. Tun Razak, therefore, tried to influence change from outside the government.

## Mobilizing Support from Outside

The two institutions that were likely to support proposals for greater intervention by the government to assist the Malay community to be more involved in the commercial and business sector of the economy were the Malay-based political parties and business associations. The open involvement of UMNO was, however, likely to cause a conflict within the Alliance and strain relations between Malay and non-Malay members in the Cabinet. Besides, Razak was not sure that he would get the support of Tunku Abdul Rahman, the Prime Minister, and other UMNO leaders. In particular, there was a danger that an open debate at the party level may strain relations between himself and the Tunku whom he knew to be against any proposal that threatened the delicate balance of power within the coalition party. His involvement could also have been construed by some members as a direct challenge to the leadership of the Tunku. For these reasons it was not considered wise to take the matter up through party channels. The other groups likely to give support to the proposals were the Malay business associations. But these associations were new and organizationally weak. They had been formed, with the assistance of the government, as a response to Chinese economic strength. This small, but increasingly assertive Malay commercial community, began to make itself heard through associational activities. In 1960, the Associated Malay Chambers of Commerce of Malaysia (AMCCM), the first national-level business association of Malay business, was formed. According to Bowie and Doner, the main objective of the association was:

> ... not so much to mobilise resources from within, but rather to press the state for more direct financial support. In the early 1960s the AMCCM brought considerable political pressure to bear on the government to allocate all government contracts worth less than M$50,000 exclusively to Malays. In response, a Royal Commission was formed in mid-1960 'to enquire into the government tender system' and although the Commission's final report failed to satisfy that particular demand, it nevertheless handed the AMCCM a victory when it recommended that 'the aims of the Government should be to allocate by administrative action not less than 25 percent of all classes of Government contracts to Malay contractors.'[10]

According to Bowie and Doner, it was the AMCCM that was instrumental in organizing the First Bumiputera Economic Congress in 1965. Although this is true in the formal sense, there is no doubt that Tun Razak had a hand in the setting up of the Congress mainly for the purpose of mobilizing support for his new policies.

It seems that the name of the Congress was carefully chosen to give it a non-political character. The term "Bumiputera" was chosen instead of "Malay", because, until then, it had no political connotation, being used as only to describe the indigenous communities of Sarawak and Sabah.

The heavy "input" of government was seen in the participants who attended the Congress. Many of them were UMNO leaders (mainly at state and local level) and government bureaucrats. From the names of the chairmen of the working committees it appears that civil servants were directly involved in the preparation of the working papers and in guiding the discussions. As to be expected, given the total absence of of non-Malays except at the opening and closing sessions, there was general agreement that government should intensify its efforts to correct the economic imbalance between the Bumiputera and the non-Bumiputera communities. Government intervention on behalf of the Bumiputera community was justified on the grounds that:

> ... in a free economy the Bumiputera is at a disadvantage because he is a late starter, and the Government should provide various types of assistance to enable a Bumiputera to engage in business or start an industry.[11]

The discussion focussed on the ways and means of government action to improve Malay participation in the business sector, particularly in the modern corporate sector. It appears that the Congress was divided on the form of assistance. While the majority wanted government to provide "more and varied assistance with the least number of conditions and restraints" to individual Malay businessmen, there was also a small group who wanted the government to take the initiative in setting up industries. A more planned approach to development was seen as necessary in order that the policy did not result in the creation of a small capitalist class while the majority of Malays failed to benefit from government assistance. Besides, it was felt that:

> ... the scale and scope of industries are getting larger and larger, and it is doubtful if local, let alone Bumiputera, entrepreneurs could undertake the setting up of new industries without a great deal of financial assistance from the Government. Certain industries may still be financed entirely by the Government if they are to be set up at all.[12]

It seems that Tun Razak was "not unsympathetic" to the view that government participation in the economy would obviate the need to provide almost unlimited financial assistance to create a Malay capitalist class.[13] But direct government involvement in the economy was associated with socialism, a political ideology that was anathema to the existing leadership. Examples from Japan were deliberately chosen to show how government intervention can be justified in order to achieve certain social and political objectives even in non-socialist countries. The working paper described how this was done during the Meiji period in Japan:

> One of the major aims of the Meiji economic policies was to transform the Samurais into modern capitalists... In many important sectors of the economy, the state started 'pioneering' factories and establishments... But as soon as the state-created enterprises became viable and began to yield profits, they were sold to private buyers, and as a rule for 15-30 percent of their actual cost. Not only were entrepreneurs allowed to purchase the factories at low price but they were also offered favourable terms of delayed payment.[14]

The resolutions passed at the Congress called for more government efforts on a number of fronts. Some of the resolutions "tallied with what the Government had already planned."[15] This was not surprising, given the input from the government bureaucracy. The recommendation for the creation of marketing boards had already been implemented with the setting up of the Federal Agricultural Marketing Authority (FAMA). Work on the reorganization of RIDA had already begun before the Congress opened. Thus, the link between the Congress and the government was very close, so close that one may even question the independence of the Congress. It appears that Congress acted as a forum for a faction within the government, headed by Tun Razak, to put pressure on the established leadership to change its policies.

At the same time, the Congress provided the opportunity for the Malay business groups to strengthen their organization in order to become a more effective pressure group and obtain more government assistance. The group was even without a secretariat or permanent staff at the time of the first Congress meeting. The Congress therefore called for the Malay Chambers of Commerce to employ qualified and experienced persons as "Secretaries to the Chambers, who should be paid initially with Government assistance."[16]

The Congress had considerable success in influencing economic policy. In fact, some of the resolutions passed at the Congress were immediately accepted by Tun Razak on behalf of the government, when he spoke at its closing session.[17] But there was no change in the general policy of government. All that happened was that existing organizations for providing various forms of assistance to Malays were strengthened and some new ones created. Another Congress was organized in 1968 to continue to press government for more assistance to Malays. At the same time criticism against the Tunku for not doing enough for the Malays, or as a corollary to that, giving too much to the non-Malays, were beginning to be expressed, albeit in sotto voce. But before the resolutions passed at the Second Bumiputera Congress could be accepted and implemented, the country was embroiled in open ethnic conflict. The Declaration of Emergency, and the suspension of Parliament that followed, resulted in Tun Razak acquiring more powers as Director of Operations, than the Prime Minister. Soon the Tunku was eased out of office and Tun Razak became the Prime Minister and the leader of UMNO. The ideas that had been expressed at the Bumiputera Congress meetings, therefore, could now be put together to form the new policy of the government.

## The Formulation of the NEP—The Core

When Tun Razak took over power one of his first tasks was to set up a special institution to formulate the new policy. The Department of National Unity (DNU) was staffed by civil servants who had worked closely with him in the Ministry of National and Rural Development. Although the functions of the DNU overlapped with the functions of the Economic Planning Unit (EPU), the unit responsible for the coordination and drawing up of economic development plans, the EPU was left out of the formulation of the new policy, at least in the initial stages.[18] It was felt that it had been too closely involved with the very policies that had now been discredited. Besides, the unit was headed by a Chinese officer and there were several non-Malay staff who, it was felt, may oppose the new policy. Barely one month after taking over office as Director of Operations, Tun Razak announced that there would be a new economic policy, but it was not at all clear how these political objectives were to be given an economic framework. The economic rationale was provided by one of the members of the team of foreign advisers who had been in the country to advise the EPU. Jus Farlan, an economic advisor from Harvard University, met secretly with the staff of the DNU and prepared a draft paper in which the government justified policies designed to achieve political objectives in terms of the broader economic and social goals of social equity and national unity. The draft was discussed and revised at a secret meeting held at Fraser's Hill which was attended by a small select group of civil servants. The draft was then submitted to Tun Razak for approval.

It was only after the policy was clearly outlined and approved by Tun Razak and the National Operations Council that it was open for public debate. The core of the policy was very similar to the proposals made in the working papers submitted at the two Bumiputera Economic Congresses. The NEP called for a more direct role for the government in the economy. The Second Malaysia Plan pointed out that:

> Direct participation by the Government in commercial and industrial undertakings represents a significant departure from past practice. The necessity for such efforts by the Government arises particularly from the aims of establishing new industrial activities in selected new growth areas and of creating a Malay commercial and industrial community.[19]

Most of all, the NEP indicated the shift in priorities of government policies. In the past, economic development plans focussed on a laissez-faire approach to economic growth. The NEP shifted the focus to a planned approach to development in which the government bureaucracy played an important role in economic decision-making. The planned approach, with an intensification of assistance programs to assist a small number of Malay capitalists, was an indication of Tun Razak's own beliefs and in his confidence in the civil service as the main agent for bringing about social and economic changes.

## The Formulation of the NEP—The Periphery

Once the main thrust of the new policy was drafted, the key decision-makers were prepared to allow wider participation in the process of its refinement. Wider participation would also ensure acceptance and cooperation, essential ingredients for the successful implementation of the policy. Non-Malay participation in the formulation of the NEP was brought about through the establishment of a multi-racial National Consultative Council (NCC). All the members of the NCC were appointed by the government. The draft proposals were discussed by the Economic Committee of the NCC for review and comments. At the meetings of the Economic Committee there was some feelings expressed that although the need to increase Malay participation in commerce and industry was recognized, there was also the need for government to continue its efforts to improve the economic position of rural Malays. It was pointed out that:

> ... Malays must have the opportunity to improve their income in their accustomed environment without being uprooted. Hence, to a very substantial extent, improvement in the Malay standard of life depended on rural development.[20]

How was this view to be reconciled with the draft proposals which emphasized Malay participation in the urban sector? According to von Vorys:

> ... few sub-committee members were inclined to oppose what appeared to be NCC policy. Draft after draft was submitted, revision after revision considered. The final version offered no integrated formula, but a simple aggregate of two separate approaches. In the traditional rural as well as the modern urban and modern rural sectors Malays were to enjoy special privileges and hold claims on extensive public resources.[21]

The two approaches were incorporated in the Report of the Economic Committee of the NCC. The Report states:

> The Committee has analysed the notion of economic imbalance in three areas, namely income, employment and ownership of wealth, and has agreed that the ways to remedy the problem lie in two main directions:
>
> (1) Intra-sectoral reforms: i.e., the level of incomes received by Malays in the different sectors, especially in the traditional rural sector, should be raised; and

(2) Inter-sectoral reforms: i.e., Malay workers should be relocated between sectors in such a way that a better redistribution, more comparable to that of non-Malay working groups, is obtained.[22]

The draft was further refined to make it more acceptable to non-Malays. On the recommendation of the Economic Planning Unit, the poverty eradication objective was extended to cover all ethnic groups[23] and in order to allay the fears of the non-Malay business community, it was emphasized that in the implementation of the policy the government will ensure that no particular group will experience any loss or feel any sense of deprivation. In his speech introducing the NEP in Parliament, Tun Abdul Razak stressed that:

> ... it appears that there are anxieties among certain quarters that its (the NEP) implementation will mean taking away from those who are now in relatively favourable positions and giving them to the have-nots. I wish to stress categorically that the Government has no such intention. The Government will be fair and the rights, properties, or privileges now belonging to whichever groups of individuals will not be taken away and be given to others. What is envisaged by the Government is that the newly created opportunities will be distributed in a just and equitable manner.[24]

It was assumed that the country would continue to enjoy a high rate of economic growth making it possible for the Malay community to get a bigger share of the benefits resulting from the higher growth rates. Policies to increase the Malay share of the economy, therefore, had to take into account the impact of these policies on the long-term economic prospects of economic growth. This meant that civil servants, charged with the responsibility of deciding economic policies in the future, had to be politically neutral and free from what may be seen as excessive political interference. Tun Razak had met with this problem as Minister of Rural Development. In order to reduce political interference at state and district levels, he had set up district and state development committees to give politicians a chance to participate in policy-making, or at least give them a feeling that they were participating in decision-making, and thus prevent them from criticizing government policies. All complaints from politicians were channelled to the higher levels of the party before being sent to the relevant department for action.

In order to protect the civil servants from politicians who may try to use the NEP for personal gain, Tun Razak established Economic Bureaus of UMNO at national and state levels. These Bureaus were to be responsible for monitoring the progress of the NEP. Complaints against the bureaucracy would be channelled through the Bureaus to the top party leadership, which would then take the matter up through the administrative channels. In this way the bureaus would act as a buffer between the administrator and the politician.

The draft proposals of the new policy were submitted to the national level Economic Bureau for comments. As expected, the Bureau accepted the proposals and suggested that in order to instill a sense of urgency to the bureaucracy, it should be stated that the policy be implemented in 20 years. This suggestion was at first not accepted by the bureaucracy. The policy was a brave attempt at changing the whole ethnic pattern of ownership of the economy, but there was as yet no clearly worked out plans for achieving the goals, and it was difficult to set a realistic time limit for achieving any physical targets. But Jus Farlan persuaded the bureaucrats that a time frame, however unrealistic, would be useful as a yardstick to measure the success of the NEP. An arbitrary target figure therefore had to be found with which to measure the success of the NEP after the 20 year period. The bureaucrats decided that if the Malay share of the corporate sector could increase from the present figure of less than two percent to 30 percent in the 20 year period, the NEP could be regarded as a success. Once this figure was accepted, the 30 percent target was arbitrarily divided into four five-year plan periods. This was then incorporated into the Second Malaysia Plan.[25]

The target, then, was not based on any realistic appraisal of the situation and the uncertainties and difficulties of achieving a policy which had such far-reaching implications for the whole society. It was merely added on to the draft proposals in order to accommodate a particular interest group. The Second Malaysia Plan states that:

The Government has set a target that within a period of 20 years, Malays and other indigenous people will manage and own at least 30 percent of the total commercial and industrial activities in all categories and scales of operation.[26]

## Key Players in Policy Formulation Since the NEP

One of the most significant consequences of the NEP was the different orientation of the government towards private economic activities. The planned development approach has as its dominant feature the setting of substantive social and economic goals. This means that the main decision-making function is carried out within the government. Although the government is influenced by pressure groups and political claimants, it is the source of all policy innovations in the system.

But within the government, there have been significant changes in the relative power of the different groups that constitute the "government". At the political level, the influence of the non-Malay parties in the Cabinet have been considerably reduced. Before the NEP was formulated, all major decisions involving matters of economic policy had to be approved by the Cabinet. But after the NEP was passed by the legislature, these decisions were made by a small group within the government who did not have to get the approval of the Cabinet as they were considered to be part of the "implementation" process. Thus the leaders of the non-Malay parties found that they were not in the same position as before to influence the policy-making process. The all-encompassing NEP gave considerable powers to the key decision-makers to decide which policies to give priority to and which to ignore. Pleas for those aspects of the NEP which benefitted non-Malays, especially the poorer sections of the Indian community, were largely ignored. But the non-Malay leaders of the political parties in the coalition were able to influence some decisions in their favour. The creation of business companies owned and controlled by all the major parties in the coalition resulted in the leaders negotiating among themselves over who was to get what. Although there was no doubt that the NEP gave a bigger share to the Malay community, and especially to UMNO leaders and supporters, the Malaysian Chinese Association (MCA) and the Malaysian Indian

Congress (MIC) leaders and supporters also got a share. At the administrative level, the MCS officers played a key role in initiating ideas and in deciding the priorities of the new policy in the initial years. The creation of a large number of government-owned or controlled public enterprises gave considerable opportunity for the officers close to Tun Razak to take over and run business operations, giving rise to a new class of "economic bureaucrats". These economic bureaucrats occupied positions of influence in the Economic Planning Unit, and in the ministries of Finance, Trade and Industry, and Public Enterprise, and together decided the economic policy of the country.

The economic policies established by this group reflected the priorities of the dominant group that had been responsible for the changes in policy in the first place. In fact, many of them were part of the group who had influenced the change of policy. These planners concentrated their efforts on finding ways and means to implement what they knew were the core policies of the NEP—the objectives of increasing the Bumiputera share of wealth in the country. At the same time, the efforts to improve the economic position of the Malays in the rural areas were continued. This group was considered politically important to UMNO, the dominant party in the ruling coalition. The other policies contained in the NEP did not receive the same priority. In particular, despite the declaration that anti-poverty programs would be implemented in all poor areas, the areas where non-Malay poor lived, mainly in the estates and in the squatter areas in the urban periphery, were neglected. Also, despite the expressed intentions of the government that the policy would be implemented in such a way so that there will not be any discrimination of any individual or group, the government decision-makers did not create the machinery to ensure that this policy was carried out.

Instead, the focus of all government efforts was on creating the Bumiputera commercial and industrial community and of achieving the target of a 30 percent Bumiputera share of the corporate sector by 1990. In trying to achieve these targets, all kinds of policies were implemented. Various regulations were introduced to force the larger private sector companies to restructure their equity ownership and employment pattern to ensure a larger Bumiputera participation. Bumiputeras were given preference in all government contracts and in employment and promotion at all levels of the public sector.

These policies opened up new opportunities for all Bumiputeras, but especially those who were well placed to take advantage of them—UMNO politicians and government bureaucrats. The involvement of the political and administrative elite in business activities was an important consequence of the NEP. In particular, political parties were allowed to undertake business activities. Although ministers and other officers holding public office were barred from engaging in business activities, this was not seriously enforced. The involvement of government officials in business operations on behalf of the government, or of the political party they represent, or even in their personal capacity, contributed to a dangerous blurring of roles and responsibilities. Conflicts of interest are bound to arise when the same person or group of persons play different roles. For example, what happens when the Minister of Finance finds that his role as the national exchequer in charge of managing the national budget, clashes with his role as Treasurer of UMNO, or his own personal business interests? The businesses owned or controlled by important politicians tend to hold a privileged relationship with the bureaucracy. The channels of preferential access are not formalized, but they exist in the informal social networks. Privileged access to government information gives a distinct advantage to a particular business company or merchant bank wishing to compete for an important government contract or bid for a project that is to be privatized. This is particularly so in a society where public access to government information is very difficult, if not impossible, to obtain.

The development orientation of government policy involves the setting of substantive social and economic goals and establishing an important role for the government in achieving these goals. This means that a different kind of government-business relationship is developed, a relationship which puts the private business sector in some kind of dependent relationship with the government. This dependent relationship works differently in the Bumiputera and non-Bumiputera business sectors. In the Bumiputera sector, the NEP has increased the opportunities for those in position of power to decide who is to benefit from the various privileges given to Bumiputeras. In the non-Bumiputera sector, the various government committees appointed to administer the rules and regulations affecting business make important decisions affecting the expansion of individual companies, mergers and take-overs and who should be involved in the equity restructuring exercises of private sector companies. In addition, loans to businesses, owned both by Bumiputera and non-Bumiputera alike, can be controlled by the same group through its control of the commercial banks. This small group is now in a better position than ever to control the private business sector, and to make policies that affect the economy as a whole. Furthermore, the policies of this group tend to coincide with the policies that the business groups would like to see implemented. Thus, whereas the NEP was seen as threatening to the business sector, the new policies, which emphasize economic growth and give importance to the private sector as the main engine of economic growth, are likely to get the support of the business sector, and the country as a whole. The emergence of a business elite, made up of managers of government corporations, multinational corporations and Malay and Chinese entrepreneurs, has resulted in the creation of "mixed" enterprises in which foreign and Bumiputera partners work together with public enterprise managers and government bureaucrats for their mutual interest. Thus, far from sabotaging government policies, the private sector has been willing to cooperate with the government. In this rather blurred division between the public and private sectors, the central institutions—that is, the UMNO party leadership, the bureaucracy and the larger business concerns—have a skewed relationship with each other. UMNO leaders protect the bureaucracy from public criticism, mainly relating to unfairness and discrimination in its decisions, while making sure that the bureaucracy carries out the policies that will ensure its continued stay in office. The bureaucracy, in turn, provides the economic justification for decisions made to serve the interests of their political masters and thus give legitimacy to the government. And the business community participates in joint ventures with UMNO and individual Bumiputera companies and expands the opportunities for economic growth and greater opportunities for the transfer of wealth to Bumiputeras.

Thus it is likely that economic policies in the future will focus much more on improving the opportunities for private businesses to expand and less on the creation and growth of public enterprises. Although the government would continue to assist the poor Malays in the rural areas, there is likely to be a reduction in the subsidies and other handouts. Policies which emphasize the efficiency of the private sector are likely to also emphasize the need for a privatization policy that covers almost all government business activities, including those activities that are natural monopolies. Privatization gives opportunities for government assets to be transferred to those especially selected to receive them, at prices well below the market value, thus further strengthening the vertical relationships of the patron-client style of leadership. Thus, while small and medium-scale companies owned by the government will be sold directly to Bumiputera individuals and companies, the larger government corporations such as the National Electricity Board, as well as the large government contracts to build and maintain the highways, are likely to be sold to the larger companies in which the ownership is made up of a combination of Bumiputera and foreign companies. The government has announced that 287 companies with capital of less than M$2.5 million are to be sold to Bumiputera investors, entrepreneurs, companies, institutions and employees of the firms concerned. The selling price of these companies would not be made known to the public but "would depend on negotiations between Bumiputera buyers and the Federal or State governments".[27] The government has also announced that foreign investors would be allowed a 25 percent equity participation when the National Electricity Board is privatized.[28]

In the past, privatization policies have also benefitted Bumiputeras who are not actively engaged in business through their participation in the Amanah Saham Nasional, a unit trust company. The transfer of profitable public enterprises to the Perbadanan Nasional Berhad (PNB), a company set up by the government to manage the privatized companies on behalf of the trust company, has resulted in a spread of the benefits of privatization to a wider section of the Bumiputera population. But it is likely that the role of PNB as the main instrument for transferring government assets to Bumiputeras will be reduced in the future. As competition for privatized projects increases, it is likely that there will be pressure for more projects earmarked for privatization to be given to private individuals and companies instead of the PNB. These pressures have taken the form of criticisms against the PNB for not contributing to the creation of "a genuine class of Bumiputera entrepreneurs". It has been suggested that:

> increasing Bumiputera ownership through expansion and growth of trust agencies such as PNB ought to be de-emphasised while priority should to be given to improving efficiency, management and productivity of existing firms in which trust agencies have a stake.[29]

Policies to encourage the growth of a genuine Bumiputera commercial and industrial class include the giving of a new set of incentives and tax holidays. These policies reflect a fundamental change in the attitude of government towards the private (business) sector. Instead of competing with the private sector, the role of government in the future will be to support private business interests. The government introduced the "Malaysia Incorporated" concept to emphasize the complementary relationship of the public and private sectors. Prime Minister Dr. Mahathir Mohammad explained the concept as follows:

> In a nation the private sector forms the commercial and economic arm of the national enterprise, while the government lays down the major policy framework and direction and provides the necessary back-up services. Thus the government becomes more the service arm of the enterprise.
>
> The Malaysia Incorporated concept, then, requires that that the economic and service arms of the nation work in full cooperation so that the nation as a whole can gain the way that a well-run corporation prospers.[30]

As the new policies are likely to benefit not only the Bumiputera business sector, but the whole business sector, albeit not to the same extent, the government decision-makers are more willing to open the debate on the new policy to a wider section of the society than at the time the NEP was formulated. Public debate on the kind of economic policy the country should have in the future is encouraged. The government has set up a special committee, the National Economic Consultative Council (NECC), composed of a wide cross-section of the society to review the NEP and make recommendations for future economic policy. In addition, numerous conferences have been organized through government-sponsored agencies to debate future policy.

It is, however, unlikely that there will be a consensus on what future policy will be. Instead, policies are likely to continue to be formulated by the same small group that make up the political and administrative elite. But this group is likely to be influenced to a large extent by private sector business interests, particularly where these interests coincide with their own business interests. The re-employment of retired bureaucrats on the boards of companies and banks that are owned, directly of indirectly, by these individuals contribute to the illusion of a "government-business" consensus. But while certain interest groups associated with particular political personalities hold a special relationship with the bureaucracy and have privileged access, non-Malay business groups may have much less access, although perhaps more than they had during the period of the NEP.

In the case of outsiders—for example, businessmen, groups associated with opposition political parties, groups hostile to the government, workers' groups and consumer groups—the government policy is to ignore them, or if they become influential, to restrict their activities through legislative action. The government has been able to ignore the demands of the workers, at least up to now, by resorting to a combination of exhortations and threats. The "Look East" policy was a policy aimed at encouraging workers to improve their productivity and sense of discipline in the work place.[31] So far the workers have remained a disparate, collectively passive force. The percentage of the population employed in the secondary sector is still small and a large component of the industrial labour force consists of females employed in the electronic industries where it is illegal to establish trade unions. But while economic policies in the future are likely to be geared towards the promotion of economic growth and the development of large-scale industries, it is likely that the new set of policies would be packaged in the same way as the NEP in order to give the consensus that is needed to get the cooperation of all groups in the society. The new policy is likely to be packaged into a comprehensive policy which incorporates the poverty eradication and restructuring goals of the NEP.

## Notes

1. James C. Scott, *Comparative Political Corruption*, (Englewoodcliffs, New Jersey: Prentice Hall Inc., 1972), p.23.

2. Robert Tilman, *Bureaucratic Transition in Malaya*, (Durham N.C.: Duke University Press, 1964), p. 102.

3. See especially Mavis Puthucheary, *The Politics of Administration: the Malaysian Experience*, (Kuala Lumpur: Oxford University Press, 1978) and Khasnor Johan, *The Emergence of the Modern Malay Administrative Elite*, (Kuala Lumpur: Oxford University Press, 1984). Although the name of the administrative service has changed, the old name, Malayan Civil Service (MCS) is retained in the text to maintain continuity.

4. Gayl Ness, *Bureaucracy and Rural Development in Malaysia*, (University of California Press, 1967), p. 226.

5. Ibid., p. 225.

6. Ibid., p. 157.

7. *Second Five-Year Plan—1961-1965*, Government Printer, 1961, pp. 50-54.

8. Gayl Ness, p. 225.

9. Wan Abdul Hamid, Manager of Small Industries Division of RIDA, in a paper entitled "Some Suggestions on the Reorganization of the Rural and Industrial Authority" dated 14 January 1965.

10. Alasdair Bowie and Richard E. Doner "Business Associations in Malaysia: Communalism and Nationalism in Organizational Growth." Paper prepared for the 1988 Annual Meeting of the American Political Science Association. Washington, D.C. (September 1-4, 1988).

11. Working Paper entitled "Participation in Industry by Bumiputera." Paper prepared for the First Bumiputera Economic Congress held in June 1965.

12. Wan Abdul Hamid, "The Bumiputera Congress—A Review" in *Sunday Mail*, 13 June 1965.

13. Interview with Dato Abdullah Ahmad, former political secretary to Tun Abdul Razak, on 13 August 1989.

14. Working paper on "Participation of Bumiputera in Industry," for the First Bumiputera Congress held in June 1965.

15. Wan Abdul Hamid in *Sunday Mail,* 13 June 1965.

16. Bumiputera Economic Congress, June 1965.

17. *Sunday Mail,* 13 June 1965.

18. Information obtained from interviews with key civil servants at the time.

19. *Second Malaysia Plan,* 1971-1975, p. 7.

20. Karl von Vorys, *Democracy without Consensus,* (Singapore: Oxford University Press, 1976), p. 405.

21. Ibid.

22. *The Report of the Economic Committee of the National Consultative Council on the Problems of Racial Economic Imbalance and National Unity,* (Kuala Lumpur: National Unity Department, Parliament Building, August 1970).

23. Interview with Thong Yah Hong, former Director-General of the Economic Planning Unit, on 5 July 1988.

24. Speech by Tun Abdul Razak, Prime Minister of Malaysia, when introducing the NEP in Parliament on 12 July 1971.

25. Interview with Dr. Agoes Salim, former Research Director in the Department of National Unity, on 4 October, 1988.

26. *Second Malaysia Plan,* p. 41.

27. *New Straits Times,* 8 July 1989.

28. *New Straits Times,* 2 August 1989.

29. See paper entitled "The Post-1990 Economic Policy: Implications for Sarawak" by Dr. Kamal Salih, the Executive Director of the Malaysian Institute of Economic Research presented at the Workshop on Development held in Kuching, October 1988, and his comments at the Conference on the economic policy after 1990 organized by the Malaysian Institute of Economic Research and the National Chamber of Commerce and Industry of Malaysia held in Kuala Lumpur on 1 August 1989 and reported in the New Straits Times of 2 August 1989.

30. *New Straits Times,* 11 October 1983.

31. See Ozay Mehmet, "Mahathir, Ataturk and Development", in *The Mahathir Era*, edited by V. Kanapathy et al., (Kuala Lumpur: International Investments Consultants, 1989), p. 36-47.

# Singapore

## Changing the Economic Policy-Making Process in Singapore: Promise and Problems

*Linda Low*

## 1. Economic Policy and the Role of Government

The focus of this paper will be on the role of the government in economic policy-making in Singapore, emphasizing in particular the change in mode. This introductory section attempts to give a broad sweep of some basic issues affecting the role of the government in a small, open city state like Singapore and how they impinge upon the relative roles of the public and private sectors (Lim and Associates, 1988, pp. 59-72). The following sections will analyse these same areas in the Singapore context, beginning with a discussion of the origins of economic planning in Singapore.[1] The evolution from formal planning to ad hoc planning, from macro-economic modelling to using leading indicators for forecasting, and from an autocratic to a more consultative and participatory style of government and economic management, will then be discussed, highlighting the institutional processes and reforms at each phase. The paper will conclude with a tentative prognosis of the roles of both economic policy-making and the government in Singapore as the baton is passed to the private sector and the privatization program commences.

The unique features of being small, resource-lacking and highly dependent on trade and the international economy, are often seen to be necessary and sufficient conditions for a fairly large role for the government. Apart from its Confucian and paternalistic tendencies, the People's Action Party (PAP), which has formed the government since 1959, has generally adopted an ideology-free, eclectic approach. Its predisposition to intervene in the economy reflects, in part, some social preferences derived from the PAP being founded on some socialist ideals.[2] In addition, the volatility of the capitalist system has also extended the scope of macro-economic planning and management. The state in Singapore has assumed a large role, as is the case in most developing economies since the Keynesian revolution.

The high level of dependence on the international economy, in terms of trade, capital and even labour inflows, has also helped to facilitate and legitimize the government's dominance. The Singapore economy is one of the most open in the world, with a total trade to gross domestic product (GDP) ratio of 369.8 percent at the peak in 1980. The ratio in 1987 declined to 307.1 percent due to the economic slowdown. The manufacturing sector in Singapore is critically dependent on foreign investment, not so much for the capital as for the technology and markets accompanying it.[3] The foreign labour supply amounts to some 12 percent of the total labour force. Like most governments with economies very dependent on external forces, Singapore has attempted to constrain the effects of the open economy on production and employment by increasing the scope of the public economy (Whiteley, 1986, pp. 47-48).

Furthermore, in the Singapore context, economic viability and political survival appear to be two sides of the same coin. Political stability is a crucial condition for economic performance just as economic success underpins political continuity. This accounts for the zealous government attempts to adopt a holistic, multi-dimensional approach to most growth and development issues. It thus offers little or no apologies for a no-holds-barred modus operandi for implementing policies when it perceives a problem. While it may be deemed arrogant for believing that whatever it has set itself to do will indeed succeed, the government's development-oriented strategies have consistently delivered the goods, in contrast to many countries in the region (Lim, 1983).

## 2. Origins of Economic Planning in Singapore

Since 1959, state involvement in economic policy-making has been significant under the PAP government, in sharp contrast to the laissez-faire approach of the British colonial government. Economic planning in Singapore was initiated as a response to a institutional demand of the World Bank in the 1960s (Schulze, 1984 and Low, 1985). In the words of the economic architect, the first Minister of Finance Dr. Goh Keng Swee, who remoulded Singapore from an entrepot economy to a viable manufacturing-cum-services city-state:

> Actually when we first won the elections in 1959, we had no plans at all. We produced a formal document called the First Four Year Plan in 1960, only because the World Bank wanted a plan. We cooked it up during a long weekend. I have very little confidence in economic planning. Planning as we know it has a limited value. Economic policy is more important. (*This Singapore*, 1975, p 34)

State planning was formalized in the first development plan for 1960 to 1964, which was extended to include 1965. The main theme in the plan was to restructure the economy away from entrepot activities toward a more dynamic manufacturing sector. Industrialization efforts had begun long before the first development plan, under earlier recommendations of the expert team from the Colombo Plan, which documented the Lyle Report of 1957. The same view was upheld by the United Nations economic adviser, Dr. A. Winsemius, whose efforts twinned with those of other pioneer PAP economic helmsmen in birthing modern Singapore.[4]

Under the first development plan, vital institutions to drive the industrialization program, such as the Economic Development Board (EDB), together with complementary social agencies like the Housing and Development Board (HDB), were launched. They took over impotent predecessors set up by the British, but still faced monumentous struggles given the economic and social pressures of those days.

The circumstances which faced freshly independent Singapore after 140 years of British colonial rule, were quite dismal. Despite the political victory over left-wing elements of the PAP, political consolidation and nation-building were onerous tasks not made easier by residual Communist agitators who took their struggle behind industrial unrest. Socially and economically, there was the equally uneasy problem of reconciling stagnating entrepot trade with a rapidly growing population demanding massive creation of job opportunities and basic social amenities. While the British left a fairly fine civil service, its achievements in the areas of education, housing, health and other social services were much to be desired.

Under such a fragile socio-political and economic scenario, Singapore sought economic merger with Malaysia, which had the immediate promise of a common market for its import-substituting industrialization strategy. It was the most rational and feasible option at that time. In 1963, the economy had lapsed into a recession and even public sector pay had to be cut. Hindsight does not disprove the logic of the merger, although perhaps it is less favourable vis-à-vis the impatience with which Singapore sought to see its vision of the Malaysian federation established. More than economics, it was sheer political and personality cross wills which brought the two-year economic union to an abrupt termination in 1965.

If Singapore's economic straits were dire when it joined Malaysia, they were even more bleak in 1965. With a common market completely struck off, its options appeared more attenuated. However, the Malaysian experiment may be deemed a blessing in disguise as it brought an urgency to switch out of import-substitution, which is inevitably a short-term measure, into an export-orientation. Once this decision was made, the PAP government went about it most assiduously and diligently, leaving no stone unturned in the drive to realize its economic goals.

Economic planning was, however, discontinued, although a second development plan for 1966 to 1970 was prepared, but never published, by the Economic Planning Unit (EPU) which did the first plan. The EPU was transferred from the Prime Minister's Office to the Ministry of Finance in 1966 to prepare the second development plan. Its efforts under foreign experts were, however, viewed quite disdainfully, partly because of the PAP government's lack of faith in planning, and partly circumstantial. Apart from the political trauma and the breakdown of the Malaysian common market, other events further changed the circumstances upon which the second plan was premised. These included the Indonesian confrontation which lasted the full duration of the Malaysian years and cost Singapore dearly in terms of entrepot trade; the British pound devaluation in 1967 which resulted in a 15 percent loss of Singapore's L200 million reserves held in British securities; and the British announcement in 1968 of a phased withdrawal of its military bases from Singapore. The latter had contributed as much as one-fifth to gross national product in the early 1960s. Besides the economic vacuum, Singapore had also to start building up its own military defence which meant a further drain on its resources.

The abortion of formal plans did not, however, spell the end of economic planning per se. In its place, Singapore embarked on a short-term, counter-recessionary plan between 1968 and 1970 to combat the anticipated recession following the British withdrawal. Within the public sector, some planning was practised behind the scene under five-year rolling plans which were not publicly released. The first of these was for the period 1972/73 to 1976/77 and was designed to serve as a medium-term plan to optimize resource allocation. Rolling plans were perceived to be more flexible and amenable to changes faced by a small, open economy, as well as being less taxing in terms of statistical and data inputs. The EPU was disbanded and its staff transferred to the Economic Development Division (EDD) of the Ministry of Finance which became the de facto administrative planning agency.

The dispensation of formal planning, opting instead for ad hoc planning, may be rationalized in five ways. First, as noted, the sudden changes in events had invalidated the second development plan. Second, inputs to planning were deemed insufficient as the statistical system, including the national income accounts, was yet to be developed. Administratively, there was also a shortage of skilled manpower, and the fusing of the EPU with the EDD was done to utilize such manpower more efficiently. Third, as events had proven, planning in a small, open economy, more dependent on vital exogenous variables such as trade and foreign investment than domestic demand factors, is extremely difficult if not impossible.

In contrast, two other factors make formal planning quite unnecessary in the context of Singapore. These include the continuity of the PAP government and its successful administrative style of combining centralization at the top level of decision-making with decentralization at the operating level. That the PAP has been given the mandate to govern election after election since 1959 means that its economic philosophy and policies can be worked into the economy without any interruption or opposition. Moreover, since the development-oriented policies actually produced tangible results, success begets more success, be it due to ad hoc or formal plans.

The mini-scale of the economy also makes unnecessary the need for formal plans to coordinate sectors and meticulously harmonize competing interests. Once the decisions are made at the highest cabinet level, which becomes the de facto dynamo for the planning agency, they are implemented by an efficient and high calibre civil service. With the main implementing agencies, including statutory boards, large, but few in numbers, cooperation rather than inter-departmental friction also ensures the success of micro-planning.

## 3. Institutional Framework and Forces in Planning

The key planning agencies, be they under the formal or ad hoc planning approach, have been the EPU under the Prime Minister's Office (1960-1966) and later under the Ministry of Finance (1966-1968); the EDD, later the Development Division of the Ministry of Finance (1968-1979); and currently, the Ministry of Trade and Industry (MTI). By 1979, a change at the economic helm had also occurred. While Dr. Goh Keng Swee left the Ministry of Finance to head other ministries (Defence and Education) and the late Mr. Hon Sue Sen was the Minister of Finance, they worked very much hand-in-glove as Mr. Hon had served as chairman of the EDB and was, with Dr. Goh, among the pioneers in economic planning.

The first Minister of MTI was Mr. Goh Chok Tong, under whom the restructuring of the economy was initiated. He was succeeded by Dr. Tony Tan who unveiled the "indicative" ten-year development plan for the 1980s. This plan was aimed at developing Singapore into a modern industrial economy based on science, technology, skills and knowledge. It is only indicative in the sense it has to be fleshed out and implemented by the executive operating ministries and statutory boards and more critically by the private sector. This indicative plan for the 1980s thus marked the shift towards more concrete inputs from the private sector but, unfortunately, the partnership was not given the opportunity to firm up.

The recession in 1985 disrupted even the indicative plan. An Economic Committee, whose chairman became the next MTI Minister, was established in April 1985 and its deliberations, culminating in its report in February 1986, formed the blueprint for the economy for the 1990s. The Economic Committee became the harbinger of collaborative efforts between public and private sectors, setting in motion a more consultative and participatory style of government. In addition to seeking the causes of the recession and remedies for it, the report of the Economic Committee also made an in-depth analysis of the economy and recharted new economic directions. Its purview was both economy-wide horizontally, as well as sector-wise laterally. The service sector is explicitly upgraded as a new growth pole just as the baton is passed to the private sector with the contraction of the public sector. While the Economic Committee appears to have completed its task, the Economic Research Unit (ERU) set up within the MTI to monitor economic trends and policies following the recession, has on-going activities.

The Monetary Authority of Singapore (MAS) also has a role in economic planning and policy-making, both directly because of it being at the apex of the financial system, as well as indirectly when Dr. Goh was its chairman. The interlocking influence of the same key personalities in different capacities and agencies cannot be understated in the Singapore context of economic policy design and implementation. It provides continuity and stability in lieu of a more formal planning mechanism.

Outside the government, the academia may be considered as the only other institution involved in assisting economic planning and policy. The Economic Research Centre (ERC) within the University of Singapore, set up with a Ford Foundation grant, was involved in manpower projections in the 1970s. Such assistance is, however, an arm's length affair as decision-making remains highly centralized behind closed doors at the top cabinet level.

This was more so before the open and consensus-seeking style of the Economic Committee which implicitly acknowledged the anachronism of top-down fait accompli policy-making.

The planning and economic decision-making process as discussed in the previous section may have been suitable for the problems and issues in the 1960s. These may be externally induced or from internal sources because of structural constraints which give little scope for more time-consuming, consensual-seeking styles. The leadership may have been in too much of a hurry, confident in itself and less interested in interacting with the less mature private sector, although that sector seemed quite willing to be led. In fact, the free market economy in Singapore has been deemed a myth given the heavy intervention in so many economic and social areas (Lim, 1983). Another view is that the economy is too small for large numbers to operate, favouring government intervention instead of competitive market forces (Krause, et al., 1987).

The circumstances which endorsed the "crisis" mentality and the approach to policy-making in knife-edge situations have changed and a more educated, younger and critical electorate now demanded a new political culture.[5] The second generation of leaders, under the tutelage of their seniors, have responded fairly well to these new demands.[6] The changeover has not, however, been entirely easy. On one hand, the old PAP guard is protective of the success of its economic policies over the years. The lynchpin of these policies has been a hyperactive public sector with either the government or its statutory boards involved in every nook and crevice of the economy. On the other hand, prolonged economic paternalism damages private sector incentives, retards entrepreneurship and may in time promote bureaucratic rigidity, red-tape and corruption. The private sector is also awakening, albeit slowly. Greater inter-industry linkages and sectoral development have increased the scope for more private sector participation, particularly in the service sector.

Some involvement of the academic community in modelling was made with the establishment of an Econometric Studies Unit (ESU) within the Department of Economics and Statistics, National University of Singapore (NUS) in 1981. The EUS was financed by a private endowment fund and its mandate is to promote the application of econometrics to economic modelling and forecasting by training both academicians and MAS officers in econometrics studies. Further collaboration between the NUS and MTI occurred in 1986 over the MTI econometric model. Simultaneously, the ESU also put together its own econometric model although, because of the involvement of the ESU model-builders in the MTI model, the distinction between the two is rather moot. Apart from this common parentage, the basic salient features of the Singapore economy are quite transparent so that it cannot really be modelled too differently.

## 4. Reforms in Economic Processes and Policy-Making

### a) Approaches

The change in approach to planning ranged from formal plans in the 1960s, to ad hoc plans in the 1970s and, finally, to the indicative plan of the 1980s. When the economic recession temporarily disrupted the indicative plan for the 1990s, the government became more introspective, realizing that some of the problems are structural in nature rather than purely cyclical. For instance, the boom in the construction sector burst in 1985 because it had been buoyed up by artificial demand caused by the release of savings from the Central Provident Fund (CPF) for both public and private housing and public policies directed at a goal of 100 percent home ownership (Lim, et al., 1986 and Low, 1987). A consultative committee from the private and public sectors was set up to investigate and make recommendations for the property market (Ministry of Finance, 1986).

The report of the Economic Committee may be deemed the informal blueprint or plan of the 1990s. Its areas of inquiry are sufficiently wide-ranging and comprehensive for it to be perceived as a new economic compass. However, its authors may protest that it was never intended to be a formal plan. Nonetheless, its findings were that some of the domestic causes for the recession include the loss of international competitiveness, over-saving and suboptimal investment in the construction industry rather than in other more productive sectors, and the corrective policies implemented reflect a change in plans.

The Economic Committee has epitomized the new directions, not only for the economy, but also for the economic decision-making process. Apart from the 12-member main committee with a majority from the private sector, it was supplemented and complemented by nine subcommittees. These consist of the subcommittees on manufacturing, services, banking and financial services, international trade, local businesses, entrepreneurship development, fiscal and financial policy, and manpower. Resource persons tapped from the public and private sectors to serve on these committees, as well as the recommendations called for and refined from observations and representations from the parties affected, were testimony to the change in policy-making style. For instance, the subcommittee on local businesses seemed to have made the occasion a free-for-all complaint session, marshalling evidence and documentary proof of cases of unfair competition by government-owned companies.

The recessionary factors identified reflect the lack of attention to the cumulative macro-economic effects of public sector dominance in the economy. The erosion in competitiveness was due inter alia, to high operating costs from statutory charges and wage costs. Wage costs rose in the three-year (1979-1981) wage correction policy of the National Wages Council (NWC). While the government is one of the tripartite parties in the NWC, its implicit role is that of "mentor", at least in setting the lead as the largest employer in implementing the wage guidelines in toto. Further intervention in the labour market is via the CPF and the Skills Development Fund (SDF) elements and the payroll tax. The government, through the mandatory CPF contribution rates which have escalated in pursuance of anti-inflationary and welfarist policies, also contributed to over-saving and over-investment in residential properties.

While the Economic Committee made recommendations to rectify the economic malaises, there is still a lack of coordination among its various recommendations. For instance, while it recommends encouraging the initiative of the private sector and endorses privatization of some public enterprises, neither it nor the Public Sector Divestment Committee looked into the linkage effects of manpower and finance implications of such recommendations. To assist private sector growth and development, the EDB has spear-headed assistance to develop local entrepreneurship, while the Trade Development Board (TDB) has set up advisory committees to help local companies export abroad. These efforts under various agencies, while enabling specialized attention to be given, could have also benefited from a more coordinated macroeconomic framework of analysis.

In fact, a shortcoming of the report of the Economic Committee as a planning blueprint is this absence of following through the policies of one agency in a particular sector, or on an economy-wide basis, to avoid or pre-empt bottlenecks or possible conflicts. Another case in point is the reduction of the CPF contribution rates to cut costs, indeed itself a commendable reversal of a "sacred cow" policy. But because the homeownership plan is ingeniously tied in lock-step with CPF funds, the effects on further depressing the property market, which is perhaps also desired as it was propping up the economy artificially, are borne by homeowners and private developers.[7] That the government cannot be imbued with such omnipresence and foresight to have the right policies and solutions simultaneously, is perhaps itself reflective of the hazards and inadequacies of planning.

## b) Tools

In 1972, a World Bank expert, Norman Hicks began formulating an economic model which was the precursor of later econometric models for planning and forecasting in the Ministry of Finance/Ministry of Trade and Industry. The 1972 model was intended to project the resource gap (export-import trade gap and investment-saving gap) at an assumed or target rate of GDP growth. The model was never used as it was based on a crude estimation method using fixed coefficients based on one year's observations. Though the model was revised in 1973, it was not ready for use.

Further revisions followed in 1975 with sectoral value-added and GDP projected by linking the growth of the economy to growth in the OECD countries. Earlier models had treated GDP growth as exogenous. A forecasting model thus evolved. With deflated variables used, projection of inflation became possible. The equations of the model were, however, still derived using simple regression techniques and some technical weaknesses of the model remained. With erratic projection results, the model was still not ready for policy simulation.

Another attempt in 1976 made more changes to the Hicks' model. By inverting the main model which gave projections of probable growth of Singapore's real GDP given OECD's growth, results which enabled the distribution of additional investment and development expenditure required to achieve a higher growth rate target, were obtained. More sophisticated estimation techniques also became possible with more statistical and manpower inputs. Local manpower trained in econometrics and systems engineering, among other technical skills, became more involved in modelling. The recession in 1974/75 gave the opportunity for the projections from the 1976 model to be subjected to some policy analysis. The forecasted GDP and sectoral value-added growth rates were also used as the basis for compiling manpower budgets. Implicit manpower budgeting was practised in 1973 when the manpower implications of a high GDP growth projected by the Public Sector Development Program, under the Development Division (formerly EDD), were attempted (Low, 1985). In 1971, a Manpower Planning Unit within the Development Division was established. The latter worked closely with the ERC in manpower forecasts. The first official manpower budget, however, appeared only in 1975 as the recession again interrupted the projections of high GDP growth rates. Such manpower plans, on a five-year rolling concept, were to give policy-makers an integrated perspective of future patterns of manpower supplies and demand.

The next revision of the economic model of the Ministry of Finance took place in 1977 which, besides updating and improving the equations in the light of more and better data, also incorporated new developments in the economy. For instance, the probable impact of large projects such as the petrochemical complex and the Raffles City project, as well as the effects of tax changes and CPF changes, were worked into the model.

Such modelling efforts to provide future economic trends for ad hoc policy analysis, economic planning and to guide policy-making, continued under the Ministry of Trade and Industry (MTI) which was set up in 1979 to relieve the Ministry of Finance of some of its functions, by hiving off the latter's Development Division. Formal planning appears to have devolved to ad hoc planning and goal-setting under a policy-making approach guided by projections from an economic/econometric model and intuitive steersmanship of seasoned politicians and Western-trained technocrats. In other words, a middle-of-the-road strategy straddling the extremities of rigid formal plans and total reliance on forecasting models appears in vogue. This mix affords an overall macro-economic overview of the economy through the various functional relationships.

A quantitative forecast using an econometric model may include several elements. It can give the direction of the movement in economic activity, and if there is an expected reversal in direction, when it will occur. A prediction may also indicate the magnitude of change and perhaps the length of time for which a movement is expected to persist. To enhance the usefulness of the macro-forecasting model, anticipation surveys and sequences of leading coincident, and lagging indicators are checked against for consistency.

In summary, the four active, broad approaches to short-term forecasting of the economy at large are time-series models, econometric models, anticipations survey, and cyclical indicators. Each of these corresponds to a particular aspect of the entire task. Thus, time series models are best equipped to exploit intensively the information contained in the past history of the single or several series to be predicted. Macro-econometric models can quantify the predominant relationships which theory suggests exist among a larger number of variables. Anticipation surveys enable estimations of aggregates of economic plans or the intentions of economic agents for variables over which these agents exercise considerable control. Finally, leading indicators signal and confirm certain recurrent business cycle events.

None of them is superior because of the paucity of generally agreed upon and successfully tested economic theories that would provide strict guidance for macro-modelling, and because of inadequacies of the available data, estimation, and surveying techniques. Yet, there are significant advantages to using each class or models or methods for the task it is best suited. In short, contrary to some partisan assertions and criticisms, the four approaches are essentially complementary rather than competitive. Suffice to say too, these planning and policy tools are more within the capabilities and responsibilities of the government than private interest groups, with the exception of the academia. Even the latter is, however, greatly handicapped by the availability of data and it has neither the statistical authority nor resources to conduct surveys.

For Singapore's growth strategy and policies, all these methods are tapped. Structural econometric models have evolved into forecasting ones. Surveys of business expectations are carried out quarterly, which give planners some indications and assessments of the private sector. The theory of planning is thus perhaps still pursued in spirit if not to the letter.

## c) Interplay of forces

Economic policy-making has progressed to a steady interplay between various state and non-state institutions. The degree and extent of such efforts varies with the economic areas and issues.[8] State agencies and influence still predominate despite the delinking and privatization attempts and larger involvement of private sector initiatives. In the labour sector, the government has withdrawn its guided consensual-seeking stances, deregulating across-the-board wage recommendations to allow market determination and a flexible wage system and participating only as an employer in the NWC. However, that the Minister for Labour is also the Secretary-General of the National Trades Union Congress still affords considerable room for subtle moral suasion among labour groups.

In capital and finance, the liberalization of CPF funds for members to access their savings for investment in real estate and approved financial assets, has vast implications for liquidity and fund mobility. This may be dovetailed with divestment under the privatization program, but the macro-economic contractionary impact of funds channelled from the CPF to the government when CPF members utilize their CPF savings to put into privatized public enterprises, deserves some attention.

The direction and magnitude of further government involvement in business and the larger issue of its philosophy of economic management in Singapore needs greater transparency to sustain the incipient partnership with the indigenous private sector. The policy of industrializing and restructuring the economy relying on foreign investment has been successful. It has brought in multi-national corporations (MNCs) with their combined package of capital, skills, technology and markets. Both their presence and public sector dominance have, however, been quite inhibitive for local entrepreneurs. Probably because of the size and impact of MNC operations, their interests and views were more considered by planners than smaller local firms. Neither the MNCs nor the local firms were directly included in economic policy-making until the Economic Committee sought feedback from them.

The greatest scope for partnership is through the new style of the EDB which attempts to rely less on "picking winners" under its priority lists of industries or at least tries to identify these after more consultations with the private sector. The EDB, under the MTI, has the monopoly of directing investment, sectoral targets and overall economic design. In conjunction with the National Computer Board (NCB), the Science Park and other agencies, it monitors and upgrades technological development. Manpower considerations are not neglected as the EDB, together with other educational and tertiary institutions, the NTUC and the NWC, sits on the Council for Professional and Technical Education (CPTE) set up in 1979 under the chairmanship of the MTI Minister.

Trade promotion and policy under the TDB also try to respond to private sector interests and needs. Increasingly, in almost every sphere, the greater involvement of the private sector, whether in consultative sessions, in a dialogue fashion, or in more formal committees is perceived. It is still too early to assess the long-term effects of these developments. Besides tapping the business community, the academia, through the NUS and the Nanyang Technological Institute (NTI), the newly-founded Institute of Policy Studies (IPS) in 1987 and other institutions, appears to be more drawn into major economic and social issues. The IPS in particular, is the government's think tank as it was set up by a government grant and managed by a government-appointed board. Its director and deputy director are seconded from the University. While not quite like the Korean Development Institute (KDI) which probably has a more direct link with the South Korean government, the IPS engages in policy-making research and discussions. Feedback to the government is either at a level of confidential submissions by the director or in open discussions with participation of civil servants and policy-makers, professionals from the private sector and academicians. Its publications also help to disseminate views and questions.

The National Agenda is the working plan of the PAP, but its coverage of major political and social issues also make it a channel for consultation and interactive play of forces. It is thus galvanized through national channels since the aspirations and goals are synonymous at both levels.

## 5. Whither Planning, Whither Government?

A full circle appears to have come about as the Economic Committee has asked:

> The wider question is the extent of the government's role in promoting economic development. Should it go beyond the traditional functions of a laissez-faire government, providing defence, law and order?... But should it start up individual businesses, using public funds, which it feels are necessary to complement the economy? Should it identify winners to support?

The Economic Committee's view is that it should not:

> ... Circumstances have now changed. The economy is larger, and the private sector is more developed.... But the government is likely to have the detailed and omniscient grasp of all sectors to identify which project to put money on, even if it knows which areas should be promoted. New investments, and with them the impetus for growth, have to be the responsibility of the private sector. (Economic Committee Report, 1986, pp 16-17)

The government has designated for itself a "back-room" role, though the four areas it has allowed its visible hand to be active, still allow its immense scope for intervention (*Straits Times*, 16 October 1986).[9] In the Singapore context, one questions whether its inherent economic and political constraints, which favour government intervention, limit the extent of reducing its size. Should efficiency rather than any ideology be the guiding principle? In any case, the government has proved itself one step ahead in thinking by initiating the privatization effort and pushing the private sector to be the engine of growth, while, at the same time, reserving for itself the last resort or safety net function.

The privatization program which took on a more formal pitch with the acceptance of the report of the Public Sector Divestment Committee in 1987, is, however, perceived to be rather superficial and slow. Of the 505 government-linked companies (GLCs) it studied, only 41 were recommended for privatization (15 for listing on the stock exchange, 9 for further privatization and 17 for total privatization). The value of shares of the 17 companies targetted for total privatization is less than 2 percent of the total value, and half of this is accounted for by one company, namely the national carrier, Singapore Airlines. Moreover, a generous ten-year timeframe for completion of the program has been put forward with no other specific details. Progress was further hampered by the October stock market crash of 1987. Statutory boards recommended for feasibility studies are still in the midst of conducting them.

The lack of urgency probably reflects the unique circumstances under which privatization is mooted, namely merely to shrink the successful public enterprise system. There is none of the usual reasons of inefficiency or fiscal drain as in other developing countries (Low in Cook and Kirkpatrick, eds, 1988, pp 259-280). Nonetheless, there are a number of outstanding problems which may have implications for future public-private sector cooperation.

One concern is the contractionary and liquidity impact of funds siphoned from the private sector back to the government when the GLCs are sold. The use of CPF funds and the careful timing of the divestments may minimize the impact somewhat and prevent any "indigestion" of the financial system. Of a more serious note is the availability of the right companies and entrepreneurs to come forth to take over the GLCs. Manpower and expertise is in tremendous shortage in the private sector. If the GLCs change hands too traumatically, with no continuity in management, their performance may be severely affected.

Another worrisome trend, in the Prime Minister's words, is that divestment may lead the government to:

... lose some of its ability to guide the economy not so much because government companies are sold off but because the general managers of these government-owned companies at present are from a tight circle of administrators who share the thinking of the policy-makers. They share the economic philosophy of the government and have a firm grasp of the rationale for various policies since they are privy to background problems so that they can interpret signals accurately, and react swiftly and flexibly. (*Business Times*, 12 June 1987)

His fears are that in the future, the senior managers would be more "removed or cut off from the inner sanctums of government and could not react as insiders." The government may then have to give them more briefings before the "hoist in the thinking behind policies". Nonetheless, some assurance may be gained from the managerial class created by the government and MNCs in the past 20 years. Moreover, what used to be pioneering enterprises have become routine business such that the private sector should be well able to manage them.

The degree of success in private sector response to the challenge given to lead the economy and test the government's faith in local entrepreneurship is difficult to ascertain. All that can be done has probably been done in terms of institutional support, finance and other technical assistance. However, after such a long period of overexposure to an expansive and paternal government, is the private sector still effective?[10] While the government may not be picking more winners in the industrial and corporate sectors to promote, it would probably do no harm to pick talents for both political and bureaucratic positions. If talents can be spotted and appropriated effectively, it would be a tremendous boost for the entire "privatization" effort, both of the GLCs and the economy at large. In all possibility, economic processes and policy-making will remain largely guided in the next ten years as the privatization program continues.

## Notes

1. It is not implied that economic decision-making is synonymous with economic planning. Economic policy as encompassed within public policy defined later, involves the

use of state power to achieve economic goals. Since Singapore started with formal planning to achieve economic goals, the origins of economic planning is thus discussed first.

2. Although the welfare system may be slanted towards attainment of social goals like old age security and health, they have important economic effects in terms of creating a contented labour force, see Low and Toh, 1985.

3. Governments may be hard pressed in dealing with highly mobile international capital and MNCs which have global resources to utilize and goals to satisfy.

4. Dr. Winsemius remained the government's economic adviser until his retirement in 1984. He, together with the first generation of leaders, were the main economic strategists behind the First Development Plan and all subsequent economic restructuring efforts. Such a continuity and constant factor in economic policy-making cannot be underscored.

5. It is claimed that in many circumstances, the government needs to act rapidly and effectively precisely because of the knife-edge, bestowing it the appearance of being "high-handed" and "government-knows-best"; see *Straits Times*, 8 November 1985 and 23 August 1985 respectively. For a discussion on the political economy approach of the PAP, see Lim and Associates, 1988, pp 59-66.

6. After the 1988 general elections, the Prime Minister, Lee Kuan Yew confirmed that he would hand over the premiership to the First Deputy Prime Minister, Goh Chok Tong in two years' time. In the interim, Lee would be the "company chairman" while Goh would be the "chief executive officer"; see *Straits Times*, 12 September 1988.

7. For an elaboration of the financing of private and public residential properties from CPF funds, see Report of the CPF Study Group, 1986, pp 49-62.

8. The political reforms as in Town Councils, Group Representative Constituencies and Government Parliamentary Committees which have implications for private-public sector participation, will not be discussed in this paper. For a brief discussion, see Quah, 1988, pp 144-150.

9. These comprise infrastructural development, training and education, development of institutions and companies to innovate and raise capital and develop ideas.

10. Krause, et al. (1987, p. 223) noted five areas where there is suggestive evidence that the current role of the government may carry risk in the future, these bring in gross savings, government entrepreneurship, absorption of talents by the public sector, public housing and secrecy in information.

## Bibliography

Cook, P. and C. Kirkpatrick, eds. *Privatization in Less Developing Countries.* (London: Wheatsheaf Books Ltd., 1988).

Krause, L.B., Koh Ai Tee and Lee (Tsao) Yuan. *The Singapore Economy Reconsidered.* (Singapore: Institute of Southeast Asian Studies, 1987).

Langford, J.W. and K.L. Brownsey, eds. *The Changing Shape of Government in Asia-Pacific Region.* (Halifax: Institute for Research on Public Policy, 1988).

Lim Chong Yah, et al. "Report of the Central Provident Fund Study Group." *Singapore Economic Review.* Special Issue, April 1986, Vol 31, No 1.

Lim Chong Yah and Associates. *Policy Options for the Singapore Economy.* (Singapore: McGraw-Hill Book Co., 1988).

Lim, Linda. "Singapore's Success: The Myth of the Free Market Economy." *Asian Survey*, Vol 23, No 6, June 1983.

_____, "State versus Market in the Rapidly Growing Economies of East and Southeast Asia." *Southeast Asia Business.* Centre for South and Southeast Asian Studies, University of Michigan, No 6, Summer, 1985.

_____, "Institutional and Policy Measures for Effective Manpower Planning: Singapore Case Study." in Regional Institute for Higher Education Development, *Manpower Planning in ASEAN Countries.* (Singapore: RIHED, 1985).

_____, "The Economics and Politics of Public Housing in Singapore." *Kajian Ekonomi Malaysia.* Vol 24, No 1, June 1987.

_____, "Privatization in Less Developed Countries: Singapore Case Study." in Cook, P. and C. Kirkpatrick, eds. *Privatization in Less Developing Countries.* (London: Wheatsheaf Books Ltd., 1988).

Quah, J. "Controlled Democracy, Political Stability and PAP Dominance: Government in Singapore." in Langford, J.W. and K.L. Brownsey, eds. *The Changing Shape of Government in the Asia-Pacific Region.* (Halifax: Institute for Research on Public Policy, 1988).

Regional Institute for Higher Education Development. *Manpower Planning in ASEAN Countries.* (Singapore: RIHED, 1985).

Schulze, D. "Development Planning in Singapore." *Indian Institute of Public Administration.* Vol 30, No 3, 1984.

Singapore, Ministry of Finance, Property Market Consultative Committee. *Action Plan for the Property Sector,* 1986.

Singapore, Ministry of Finance. *Report of the Public Sector Divestment Committee,* 1987.

Singapore, Ministry of Trade and Industry. Report of the Economic Committee, *The Singapore Economy: New Directions,* February 1986.

Times Publication. *This Singapore,* 1975.

Toh Mun Heng and Linda Low. "Theoretical and Empirical Issues in Social Security." in Amina Tyabji, ed. *Social Security Systems in ASEAN, ASEAN Economic Bulletin.* Special Issue, Vol 3, No 1, July 1985.

Amina Tyabji. ed, *Social Security Systems in ASEAN, ASEAN Economic Bulletin.* Special Issue, Vol 3, No 1, July 1985.

Whiteley, P. *Political Control of the Macroeconomy: The Political Economy of Public Policy Making.* (London: Sage Publications, 1986).

# Indonesia

## The Indonesian Deregulation Process: Problems, Constraints and Prospects

*Sjahrir*

### Introduction

This paper attempts to explain why deregulation has become the most important economic policy in Indonesia since 1983. It begins with a discussion of the political and institutional dimensions of economic policy-making in Indonesia. This is followed by an analysis of the deregulation policy from both theoretical and empirical perspectives. The paper then examines deregulation as a policy problem from a political economy analysis, emphasizing the constraints within which it must operate in Indonesia. Finally, the prospect of deregulation in the immediate and mid-term future is assessed.

The New Order Regime in Indonesia (from 1967 to the present) has witnessed two phases of economic growth. The high growth of 1967/1968 to 1981 was signified by economic growth at seven to eight percent per annum and an increase of income per capita by five percent or more. This phase was also marked by booming oil exports, infusion of foreign capital through foreign aid and investment, and increased production in agriculture and industry.

From 1982 to 1987 Indonesian economic growth was reduced significantly, with an average annual growth rate of approximately 3.5 percent. One contributing factor which would account for the different growth rates of the two periods is the effect of changing oil prices. The oil booms of 1973/1974 and 1980/1981 affected the Indonesian balance of payments and budget tremendously. At the peak of the oil boom (1980/1981), oil and natural gas accounted for more than 70 percent of Indonesia's exports and more than 60 percent of the Indonesian budget came from oil and gas export taxes. Similarly, the oil price drop, which started in 1982, continued in 1983 and reached its lowest level in 1986, had a major affect on Indonesia's balance of payments and budget and devaluation measures had to be taken in early 1983 and at the end of 1986. Despite these troubles, many observers were surprised at the resilience of the Indonesian economy. Other oil producing countries like Nigeria (an OPEC member) and Mexico (a non-OPEC member) have gotten into much worse economic trouble.

One other factor which became a constraint to Indonesian economic growth was the changing nature of foreign aid and investment. In the first phase of its economic growth, Indonesia received a substantial inflow of these resources. However, in the second phase she received much less and, in fact, during the last two budget years, the foreign aid effect has produced a "net resource outflow", meaning that amortization and interest payments in these budgets were larger than what was received through new foreign aid. As for foreign investment, Indonesia saw much less of it because of the changing environment for import substitution industries after the oil-boom era. There is currently much less effective demand for import substitution industries because the drop in oil prices has reduced government, as well as private expenditure.

Given the tremendous impact of the oil price drop on the economy, and understanding the changing nature of foreign aid within the budget, an obvious question is how Indonesia can still manage her foreign debt (up to the present there has been no debt rescheduling program), while at the same time realize an increase in per capita income, albeit at a low level of one to two percent per annum. The answer to that question possibly lies in the series of deregulation measures conducted by the Indonesian government from mid-1983 to the latest measure, in November 1988.

## The Importance of Deregulation

The Indonesian economy before the oil price drop was an economy in which the government was omnipotent and omnipresent. The processes of allocation, accumulation and distribution in the economy were nearly all conducted by the government. State owned enterprises were much stronger than private enterprises in nearly every sector of the economy.

The idea of the centrality of the government in the national economy has firm roots in Indonesia's constitution and there is still strong agreement that the state's role in the economy should be predominant. In banking for instance, the Indonesian Central Bank is strong not only because of its influence on policy, but also because state banks provide around 70 percent of the nation's total credit.[1] In sectors such as agriculture, or more precisely rice production, the Indonesian government's role is no less important than the government in a socialist economy.[2] Even though rice farms are not owned by the state, government influence is exercised through several policy instruments. Through the state banks, especially BRI (Bank Rakyat Indonesia), the Indonesian government provides credit facilities known as the "Bimas and Inmas package".[3] Through the budget, the government provides subsidies for the farmers to facilitate their use of insecticides and pesticides. The Ministry of Agriculture provides personnel to help farmers in the implementation of the intensification program. The State Logistics Bureau, BULOG (Badan Urusan Logistik), manages rice commodity price levels with instruments consisting of stock provision, market operations, and the setting of floor and ceiling prices, respectively protecting the farmer and the consumer. As a result of such heavy government interference, Indonesia succeeded in achieving self sufficiency in rice production in 1984. This achievement has been highly appreciated by the FAO (Food and Agriculture Organization) which gave an award to President Soeharto in recognition of it. But this production orientation and government interference has economic costs. The price of Indonesian rice is higher than that of imported rice, and

Indonesians consume more and more rice, although this may not necessarily be the most economical consumption behaviour. Regions that once had other main staple foods have switched to rice and, with the population increase and diminishing money to be spent for subsidies and credit, it will become much more difficult to maintain future self sufficiency in rice production.

The same government influence can also be seen in the growth and development of manufacturing industries. Import substitution industries were developed with a high degree of protection including import bans or quantitative restrictions, provision of subsidies and credit, and government procurement of durable consumer goods as well as capital and intermediate goods.[4]

One could go on to still other sectors, such as trade and services, but a similar picture again arises. Government economic institutions in banking, trade, insurance and re-insurance, in terms of number of activities, and in value (assets), are much larger than private institutions.

From this description of the supply side of the economy (production) and an appreciation of the government's great power (in large part a result of oil money and foreign aid) in the demand side, it is no wonder that there is very little incentive for private initiative. The government budget increased from 12 percent of the GDP in 1968 to more than 26 percent in 1981/1982.[5] This growth was largely due to the increased share of so-called "development expenditure" compared to "routine expenditure". The oil price drop changed all that, and very soon, not only was the routine expenditure share larger than that of development expenditure. For example, government procurement and expenditure in the construction industry was substantially reduced. In 1981 the government helped to produce a 22 percent growth rate in that sector. But since the drop in oil prices, there has been virtually no government expenditure in the construction sector and it posted close to a negative growth in 1986. At the same time, the foreign debt expenditure post became the largest (53 percent) post in the routine expenditure side of the budget.[6]

The fragile economic situation resulting from the oil price drop was shown very explicitly in the foreign balance of payments and budget. There existed the need for increased non-oil and gas exports to compensate for the decline in foreign exchange earnings that accompanied the drop in oil prices. There also existed the need for higher government earnings from non-oil and gas taxes to substitute for earnings from tax on oil and gas. In both cases policies aimed to fulfill these two big needs involved supply and demand changes in the economy and, because in both cases the government had to rely on the private sector, the only policy measure available was deregulation. But, before examining the deregulation policy process itself, we will briefly discuss the nature of economic policy-making in Indonesia.

## Economic Policy-Making: Political and Institutional Dimension

Like many other developing countries, economic rationality in policy decision lies with technocrats in power. In Indonesia, technocrats come mainly from the Department of Economics of the University of Indonesia. They are in charge of most of the economic posts such as the Ministries of Planning, Finance, Trade, and the Coordinating Ministry for Economic, Finance and Industry.

In 1967 President Soeharto came to power with the support of the military. He has been duly re-elected in the 1971, 1978, 1983 and 1988 general elections. His present term will end in 1993, at which time he will be 73 years old. Throughout his rule, technocrats have played a key role. There are three different periods, stemming from different economic conditions, in which the technocrats have worked with different policy prescriptions.

The first period is called the *Stabilization* period (1967-1973). During this period the main task of the technocrats was to reverse the economic disaster made by the previous Soekarno government. Rampant inflation (close to 650 percent in 1966), declining national productivity and a worsening balance of trade were the nature of the economic scene. This period was the most difficult for the technocrats. They had to control inflation, make exchange rates more realistic and renegotiate foreign debt at the same time they were applying for new debt to provide for import needs.

The second period is the *Oil Boom* period from 1974 to 1981. This was a period where oil money flew to the budget and monetary sectors through credits. Although fiscal prudence was still evident, the "balanced budget" was in balance at a higher level every year. Industrialization was developed with import substitution type of activities due to the availability of oil money. In this period, the increasing importance of the "engineer" faction inside the cabinet was pronounced. The Ministry of Research and Technology and the Ministry of Industry were held by the engineers. Huge projects such as production of air planes and ships were launched, as well as the establishment of highly capital-intensive industries such as steel and petro-chemical. Some of these industries were economically justified because they were resource based; the rest, although technologically feasible, were not. The engineers' policy of emphasizing huge projects was to eventually come in to conflict with the technocrats' *deregulation* process.

Overall, the first and the second periods produced high economic growth, but, when oil prices declined in 1982/83 and again in 1986, the deregulation policy became a necessity.

The third period is thus the *Deregulation* period. In this period, the Indonesian economy has been structurally transformed, leading to a higher per capita income. The number of people who earn their living from agriculture has significantly decreased, while wage labourers entering the manufacturing sector and services industries has rapidly increased. At the same time, the government has less money to offer in public investment due to the decline in oil prices; the impact of the foreign aid and investment, that gave Indonesia ample money to invest in social overhead capital, becomes very different in nature as, for the last three budget years, the government has had to pay more foreign debt and interest payment than it received in aid; and the industrial sector that must compete at the international level (the latest credo is shifting "oil export" to "non-oil export") has had various problems in changing its focus from import substitution to exports.

The changes in these three periods has set the stage for a struggle between forces for deregulation and those against it. In the background, which also becomes a binding constraint, is the fact that Indonesia is governed by a President that has been in power for 22 years. This constraint is significant in that the head of state is less than inclined to be a leading advocate of deregulation. Sometimes, the engineers get the upper hand and sometimes industrialists needing protection get their will. The forces for deregulation are mainly from technocrats and entrepreneurs who have minimal access to the power structure. At the international level, the World Bank and IMF are the external institutions who favour deregulation in order to reduce distortion in the economy.

Since the stage has been set and players have been identified, we are now ready to look deeper into the deregulation policy itself.

## Deregulation as a Policy Process

Deregulation is not a once and for all policy decree. It is an ongoing process that is aimed at encouraging private initiative and enhancing national growth. The problem is in defining the concept of deregulation and, as a consequence, measuring the level of change or improvement to the national economy. Hollis Chenery stated that "to achieve a given effect on production, or use of any commodity, there is a choice between controlling a price and controlling a quantity".[7] His perception of government policy effect could be used to help define the concept of deregulation. It is wrong to regard the deregulation policy process as a "dogma" or "rigid ideology" since, as a pure classical economist would point out, every government interference by definition creates a distortion in the economy. Although a policy of total non-government intervention in the economy would be impossible for a developing country to pursue, given the huge government presence in the economy and its impact of producing a "crowding out" effect on the private sector, it would probably not be too extreme for the Indonesian government to emphasize deregulation as a remedy for the problems of the post-oil boom economy. The definition that I shall use to evaluate measures or policies considered as deregulation are as follows:

1. If the government switches its policy instrument from "quantity" variable to "price" variable;

2. If the price variable interference has been reduced to a lower level;

3. If regulation became much more uniform or universal in its interpretation and less open to particular officials' interpretation; and

4. If the number of regulations have been reduced.

Given this somewhat general or loose definition, I believe it is possible to measure the deregulation process undertaken in Indonesia from 1983 to 1988. It is of course much more difficult to establish the impact of deregulation policy on economic growth, export increase and increased tax receipts because factors such as currency realignment (the two devaluations of 1983 and 1986), and the fluctuation of international prices of export goods, have had a share in explaining the economic growth that has occurred despite the oil price decline. Nevertheless, observing the deregulation policy process is still important as it helps to understand the nature of the Indonesian post-oil boom economy.

## The Deregulation Policy Process of 1983–1988

There have been many deregulation measures undertaken throughout this period which encompasses the Fourth Development Cabinet (March 1983 to March 1988) and the Fifth Development Cabinet (March 1988 to present). In both cabinets the influence of economists have been quite significant, as revealed in the deregulation policy process. For present purposes, I have taken ten coordinated policy packages which were all part of the deregulation process; a process which I think will continue in the very near future.

These deregulation measures are as follows:

1. **The first banking deregulation of June 1, 1983.** This measure gave state banks more room to manoeuver and, at the same time, reduced Central Bank interference in direct credit implementation. The quantity variable (credit reserved for state banks) was switched to a price variable (allowing state banks to determine their deposit and lending rate).

2. **The new tax laws of January 1st, 1984 on income tax and April 1st, 1985 on value added-tax.** These tax laws represent a simplification and clarification of previous laws. Tax rates were limited to only three rates, with the highest rate set at 35 percent. Tax players would also pay through a

system of "self assessment", reducing official interference in the tax system.

3. **Presidential Instruction No. 5, 1984.** This measure gave departments and high government institutions legal grounds for further deregulation, especially in licensing for investment, production and other forms of government and private sector relationships. As a result of the Presidential Instruction, the Departments of Industry, Forestry, and Trade, as well as the Central Bank, reduced licensing and procedural requirements related to private business activities.

4. **Presidential Instruction No. 4, 1985.** This has been the most substantial deregulation measure taken so far. Customs responsibility for international trade was shifted from the Directorate General of Customs and Excise of the Department of Finance, to an international private Swiss-based firm, SGS. With this measure, what many regarded as corrupt activities on the part of the Directorate General of Customs and Excise vanished almost overnight.

5. **The May 6th, 1986 Package.** This package was formulated by the Coordinating Minister for Economic, Financial, and Industrial Affairs and mainly involved a decree by the Ministers of Finance and Trade which opened up the possibility for manufacturing exporters to import goods required in the production process. Prior to this, the import of many goods (capital and raw materials) were carried out with quantitative restrictions (QRs). With this package, though quantitative restrictions still exist, exporters needing imported goods are free to import (subject to tariff), without going through licensed importers operating within QRs or non-tariff barriers (NTBs).

6. **The October 25th, 1986 Package.** For the first time the government "succeeded" in reducing QRs and NTBs for several import commodities by substituting tariff barriers. But, since the decision came only slightly more than a month after the currency devaluation of September 12, 1986, and because the number of commodities subject to change from quantity variable (NTB or QR) to price variable (tariff) were few, the public appeared somewhat dissatisfied, claiming the

measure as "too little and too late". This measure was followed by political developments as an evening newspaper, *Sinar Harapan*, was banned for printing a draft of the deregulation measure, and economic developments (the buying of U.S. dollars even after the devaluation) that led to a more significant policy measure.

7. **The January 15, 1987 Package.** This package changed the quantity variable to price variable for many more imported commodities than the October 25, 1986 package and somewhat reduced the feeling of uncertainty about the deregulation momentum.

8. **The December 24, 1987 Package.** This package continued the switch to tariffs, but also stipulated a reduction in the number of licenses. For instance, prior to this package, building a hotel required more than 25 permits, but after December 24, 1987 it required only two permits although they were now of longer duration—five years compared to one year.

In March 1988, the cabinet that initiated the deregulation policy process was replaced by a new one. The Fifth Development Cabinet continued its predecessor's policy of deregulation by producing two "packages".

9. **The October 27th, 1988 Package on the financial sector and financial institutions.** This measure substantially liberalized the financial sector of the economy. Although since 1967, Indonesia has consistently held to a "Free Foreign Exchange Regime", this latest deregulation package totally opened up the country to foreign capital and financial institutions. For example, previously new entry into the banking business was closed, which made the purchase of banking licences a known activity. However, this package allowed new banks to open with paid up capital of Rp. 10 billion, or less than US$ 6 million. Existing foreign banks could open new branches in six cities, while new foreign banks could enter the banking business in Indonesia either independently or with joint ventures between other foreign banks or domestic private banks. In addition, state banks could not monopolize deposits of state-owned enterprise,

because the enterprises were now permitted to put 50 percent of their funds in private banks.

10. **The November 21st, 1988 Package on trade, shipping, and the industrial and agricultural sectors.** This package is the most sweeping NTB reduction so far as it includes imports of plastics, a highly sensitive commodity involving a monopoly by a state-owned enterprise, in cooperation with a private firm, that has long been assessed as being engaged in "rent seeking activity" because of its monopoly position. In the shipping sector, the deregulation was also substantial because it abolished regulation of shipping lanes which previously gave officials (at The Directorate General of Sea Transportation of The Ministry of Transportation) power to distribute, some times arbitrarily, shipping lanes.

Having described the 10 deregulation measures it is now important to observe problems relating to the implementation of this policy.

## Problems of Deregulation

It is important to make a distinction between economic forces that benefit from regulation and forces that are left behind because of it. The latter will be called deregulation forces. But who are the regulation forces? And why are they for regulation? The regulation forces consist of bureaucrats who benefit from having the power to give licenses; part of the private sector protected by regulation, that is, economic forces protected from competition by import bans; and persons in power, who because of regulation, have access to goods and services that may be distributed to other persons or political factions that aid their political survival. These groups are for regulation because the regulatory market or captive markets put them in a position of economic strength that may be used for their own accumulation of capital and allows them to allocate and distribute resources according to their will, often solely to enhance their economic and political strength.

This is not a case of Indonesian uniqueness, and in fact happens not only in other developing countries, but also in developed countries. But what does possibly make the Indonesian case unique is the tremendous strength of the regulation forces which derives from the "omnipresence" and "omnipotence" of the state. It is very difficult to make a sharp distinction between persons in power and the government itself.

What is again unique in New Order Indonesia is that political stability has been achieved with less harshness and violence than that experienced in many developing countries. For 22 years and four general elections in which the state party has always been assured of victory, Indonesia has been ruled by President Soeharto and has succeeded in achieving economic growth. What has this got to do with the regulation forces? It is beyond doubt that oil money, economic regulation, and the political astuteness of President Soeharto have played an important role in contributing to political stability in Indonesia. Oil money and government intervention helped to achieve economic growth and the provision of basic human needs in Indonesia.[8] This would explain the relative stability and increased government role in the economy. Not only was the budget increased as a percentage of GDP, but state bank credit accounted for 70 percent of total credit, most of which went to state-owned enterprises. The inter-connectedness between state political and state economic institutions, the difficulty in making a distinction between persons in government and the government itself, combined with the absence of an opposition and the very minuscule size and influence of the middle class, all pointed to the direction of increased state power, which, in most part, is synonymous with regulation forces.

But another important political and economic development is the rise of the capitalist force in Indonesia[9] and the process of industrialization that went along with it.[10] It succeeded, with the help of oil money channelled through the budget and through credit, in creating an import substitution type of manufacturing sector. Later on, some industries, such as textiles (with the help of a U.S. quota, which served the interests of government officials who distributed it) were geared for export, but comprehensively changing the nature of industry to a more competitive position has been difficult because it must start with an attempt to reduce protectionism and challenge the regulation forces. That is why it was not until October 25, 1986 that NTBs were included as part of the deregulation process, and it was only the November 21, 1988 measure that resulted in their substantial reduction, that is, more than two years after the first measure to reduce them.

An obvious question that arises is why did the forces protected by NTBs eventually give in? In fact, they didn't actually give in easily, and with the new "battle cry" of non-oil and gas exports, there now exists a new QR known as the "export ban policy". The policy derives from the belief that higher value-added exports should be the aim of the future. As every businessman would be better off selling higher value-added goods, the problem is not the idea but how to achieve it. However, like NTBs on imports, this export ban is really a manipulation of quantity as opposed to a manipulation of price (through taxes and tariffs) and, as such, is contrary to the deregulation policy.

Clearly there are strong problems that derive from the state's regulating of the economy that hamper the continuation of the deregulation drive. If, despite strong international pressure to reduce protectionist measures, import restrictions are replaced by export bans, then this shows that there are always forces, within and outside the state, who benefit from regulation and will try to continue to maintain economic regulation.

## Further Constraints for Continued Deregulation

The above problems are constraints on the deregulation process, but they are not the only constraints and one must consider what may be called an "historical" constraint. The Indonesian Founding Fathers were preoccupied first and foremost with social and economic justice. Not only is this reflected in the Constitution, especially Article 33, but it has also imbued the thinking of the political elite in Indonesia from the early 20th Century to the present time. There is virtually no political or economic literature that could be categorized as being in favour of capitalism, or even competition, for that matter.

The latest deregulation measures (the October 27, 1988 package and the November 21, 1988 package) were formulated in such a way as to prevent the labelling of those measures as "liberal" and leading to "free-fight competition"; but instead, talk of the implementation of "economic democracy" and "participation in development". This distinction might not strike many people as important, but it is, because the very nature of the defence made for regulation, which is anti-liberalization of the economy, is in itself a constraint for future deregulation policy. This is so because the vague accusation of liberalization has strong political roots, not only within the state, but most importantly, within the political elite.

If Thailand, for instance, has been able to manage the problem of the position of Chinese businessman in the economy, this has not been the case with Indonesia. The Chinese Indonesians can only strengthen their position through "political, client-businessmen" relationships,[11] which in many cases, are carried out through regulation by state personnel aimed at protecting business activities of the Chinese Indonesians, as a "favour" from the state.

Meanwhile, every attempt to further deregulate the economy, and this is still important in many sectors, has to be very careful not to invite negative responses from the political public. It is of the utmost importance for the technocrats (who have become the source of the economic policy-making process) to defend their position without giving in to accusations from the regulation forces. In this context it is important to define the word efficiency, or the term economic efficiency, so as to have a strategic meaning in promoting economic growth. Even when using words like economic efficiency, one must remember the obsession of the political public or political elite with social justice or even economic socialism which constitute a strong and quite permanent constraint for continuing the deregulation policy. Accordingly, a completely deregulated economy is virtually impossible.

## The Political Economy of the Indonesian Deregulation Policy Process

State-owned enterprises play a major role in the Indonesian economy. President Soeharto, in his address to cabinet December 1986, stated that an attempt should be made to privatize state enterprises that show losses in their business activities. But as of this writing, the deregulation or privatization of state-owned enterprise has not been conducted. The history and strength of state-owned enterprises did not begin with the New Order Era but is a part of a longer Indonesian economic history. During the late President Soekarno's reign, state-owned enterprises played an important role, and had strong links with the military. (The role of retired Lieutenant General Ibnu Sutowo in the oil industry began in the Soekarno era.) Also, large trade and industrial companies owned by the state started thirty years ago.

Combined, the parts of this paper on the problems and constraints of deregulation, and on state-owned enterprises, might be presented as "the political economy of deregulation." To evaluate the deregulation policy process it is useful to define what really is meant by the political economy perspective developed in this paper.

The political economy perspective is a tool to evaluate the deregulation policy process by linking the criteria used to define deregulation with three other factors, namely:

1. The political and economic position and interaction of regulation and deregulation forces;

2. The political elite's and political public's knowledge, acceptance and resistance of deregulation; and

3. The role of state-owned enterprises in particular, and economic institutions in general, and their relation to deregulation policy.

Deregulation has been defined as a reduction in the number of licenses, the clarification of the interpretation of regulations, and the switching from the manipulation of policy from quantity variable to price variable. It is clear that the regulation forces, consisting partly of government officials and businessmen taking advantage of the distorted market resulting from regulation, will be against deregulation policy. The deregulation forces, on the other hand, are actually small in number. One would include among them, the technocrats in the cabinet and bureaucracy, intellectuals and private economists concerned about the general nature of the market for goods and services, and businessmen being unfairly excluded because of regulation. One advantage that this small deregulation force have is their possession of articulate ways of expressing their thoughts, conveyed through newspaper and news-magazine articles. Another factor working to their advantage is the pro-deregulation positions of the international financial and trade institutions, such as the International Monetary Fund, the World Bank, and GATT, that must always be taken into account by the Indonesian government because of the importance of foreign aid to the Indonesian economy. The debt service ratio for 1988 is assessed at 35 percent by the government[12] and close to 40 percent by many private economists.[13]

Evaluating the impact of the ten deregulation packages achieved so far, it would not be wrong to state that the regulation forces have been weakened somewhat. However, their voice in the political power structure appears to be steady, as proven by the increasing use of export bans for value-added policies in manufacturing.

Looking at the political elite and the political public, there is far from widespread acceptance of deregulation. Some members of parliament have already reacted by stating that the latest deregulation packages have the potential to make the Indonesian economy a liberal one. This of course does not mean that deregulation could be stopped at this stage. At present there are simply not enough resources at the government's disposal to increase growth through government investment and expenditure. Thus, since the oil price drop, along with the necessity to deregulate, there is also the necessity of increasing the value of taxes, together with a broadening of the tax base. But at the same time, there is a limit to deregulation because of the reluctance of the political elite to accept capitalism and the role of Indonesian Chinese businessmen in the economy. The problem then is to assess the level of acceptance of the deregulation measures taken so far. It appears that there are obstacles in understanding the actual level of acceptance among the political public. For example, The Indonesian Chamber of Commerce (KADIN) appears to accept the deregulation measures more enthusiastically than the government controlled political party (GOLKAR). Although GOLKAR appears to formally accept deregulation, it is clear that its officials think that the government's role in the economy will always, and must always, be important. This would exclude the desirability of a capitalist economy.

With regard to economic institutions, it is clear, especially for state-owned enterprises, that their power is mainly based on licenses and the manipulation of the quantity of goods and services. The difficulty in deregulating such economic institutions to produce a more competitive market lies in the strongly held belief in the constitution, and more importantly in the way that it is implemented to emphasize the growth and development of cooperatives. In order to make these institutions grow in number, cooperatives receive government authority, through the National Logistics Board (BULOG/Badan Urusan Logistik), to become its extension in buying rice from the farmer. So, the rise of cooperatives in the villages called KUD (Koperasi Unit Desa) is strongly linked to the distribution network of this agency. Looking at state-owned enterprises, such as the oil company PERTAMINA, the national airline GARUDA, and many plantations exporting such primary commodities as rubber, coffee, tea and palm oil, it is indeed a tall order to achieve deregulation over such tremendous economic powers. But, the November 21, 1988 deregulation package has produced a more competitive sea transportation system by reducing the power of the Directorate General of Sea Transportation within the Ministry of Transportation. Now sea lanes will be open for competition and not decided by regulation. This will affect state-owned sea transportation companies like PELNI and DJAKARTA LLOYD.

## The Prospect for Deregulation in the Future

It is clear that Indonesia cannot continue to rely on the revenue from its oil imports. In 1987, Indonesian non-oil and gas exports, for the first time in the last sixteen years, surpassed the value of oil and gas exports. This has been a result, not only of the oil price drop, but more importantly, of the increase in the value of exports. The Indonesian government's target for non-oil and gas exports in 1988/1989 is US$11.3 billion. In June, July and August 1988, average monthly non-oil and gas exports reached slightly above US$1 billion. No observer of the Indonesian economy doubts that the increase in non-oil and gas exports stems mainly from the deregulation drive, although currency realignments certainly helped to increase Indonesian exports, in dollar terms, to countries such as Japan and the European Community members, whose currencies appreciated to the U.S. dollar. But, the Indonesian economy still has to deregulate a great deal in order to compensate fully for the decline of oil prices. Not only should non-oil and gas exports be increased, but taxes should also be the main source of government budget revenue. Here, it is of the utmost importance to have a strong business community that not only has the capacity to export, but to create employment as well.

One key factor for future deregulation is the transformation of the manufacturing industry from a protective import substitution type to an export orientation. This does not mean using export bans as a policy instrument. Means of calculating the domestic use of material for production such as Domestic Resource Cost must become a yardstick for the development of the manufacturing industry. This would dictate the shift from export bans to export tariffs.

The above appraisal of economic institutions and of the political elite shows that deregulation should be sold not only as an economic measure but also as a "political good". This means that technocrats or other government officials who want to further the deregulation process must be able to sell the idea politically. In turn, this would require increased information outflow and even more articulate conveying of the meaning of deregulation and its benefits for the economy as a whole, while explaining the success of the export drive in a more systematic and appealing way.

Unless the government succeeds in selling deregulation as a political good, the political public will have a hard time accepting it. But another probably more difficult problem, is the interrelatedness of regulation forces and the political power structure. In the end, if a choice must be made between further deregulation and the maintenance of the power structure, then deregulation might be sacrificed. But, on the other hand, if external conditions such as increased foreign debt, oil price decline, and world recession create pressures for further deregulation, then the clash between economic need and political factors will become more intense and heated.

## Notes

1. Anwar Nasution, *Financial Institutions and Policies in Indonesia*, (Singapore: Institute of Southeast Asian Studies, 1983), pp 1-25.

2. C. Peter Timmer, Walter P. Falcon and Scott R. Pearson, *Food Policy Analysis*, (The Johns Hopkins University Press, 1983), Chapters 5 and 6. Also, C. Peter Timmer, *Getting Prices Right—The Scope and Limits of Agricultural Price Policy*, (Cornell University Press, 1986).

3. Ibid.

4. Hal Hill, *Foreign Investment and Industrialization in Indonesia*, (Oxford University Press, 1988).

5. See various issues, *Pidato Pertanggung Jawaban Presiden di depan DPR*, January 1980-1982/83, Departemen Penerangan.

6. President Budget Year 1988/1989.

7. From Gerald M. Meier, ed., *Leading Issues in Economic Development*, 3rd edition, (New York: Oxford University Press, 1976), pp 809-818.

8. See Sjahrir, *Basic Needs in Indonesia: Economics, Polictics and Public Policy*, (Singapore: Institute of Southeast Asian Studies, 1986).

9. See Richard Robison, *Indonesia: The Rise of Capital*, (Sydney: Allen and Unwin, 1986).

10. Hal Hill.

11. See Jahja M. Muhaimin, MIT Ph.D Thesis, *The Politics of Client-Businessmen*.

12. President Soeharto presenting budget to parliament, early January 1988.

13. See various newspapers, among others *Kompas, Bisnis Indonesia, Suara Karya*; also, see World Bank country report on Indonesia, June 1988.

# Related Publications
# January 1990

Order address:

The Institute for Research on Public Policy
P.O. Box 3670 South
Halifax, Nova Scotia
B3J 3K6
1-800-565-0659 (toll free)

| | |
|---|---|
| Keith A.J. Hay (ed.) | *Canadian Perspectives on Economic Relations With Japan.* 1980 $18.95 |
| Roy A. Matthews | *Canada and the "Little Dragons": An Analysis of Economic Developments in Hong Kong, Taiwan, and South Korea and the Challenge/Opportunity They Present for Canadian Interests in the 1980s.* 1983 $11.95 |
| Yoshi Tsurumi with Rebecca R. Tsurumi | *Sogoshosha: Engines of Export-Based Growth.* (Revised Edition). 1984 $10.95 |
| Richard W. Wright | *Japanese Business in Canada: The Elusive Alliance.* 1984 $12.00 |

| | |
|---|---|
| Conference Papers | *Canada and International Trade. Volume One: Major Issues of Canadian Trade Policy. Volume Two: Canada and the Pacific Rim.* 1985 $25.00 (set) |
| Richard W. Wright with Susan Huggett | *A Yen for Profit: Canadian Financial Institutions in Japan.* 1987 $15.00 |
| John W. Langford & K. Lorne Brownsey (eds.) | *The Changing Shape of Government in the Asia-Pacific Region.* 1988. $22.00 |
| Zhang Peiji & Ralph W. Huenemann (eds.) | *China's Foreign Trade.* 1988 $12.95 |
| K. Lorne Brownsey (ed.) | *Canada–Japan: Policy Issues for the Future.* 1989 $19.95 |

## DATE DUE

MAY 12 1995

Demco, Inc. 38-293